Authors and Reviewers

Bill Jackson
Jenny Kempe
Tricia Salerno
Dr. Leslie Arceneaux
Allison Coates

Teacher's Guide

Published by Singapore Math Inc.

19535 SW 129th Avenue
Tualatin, OR 97062
www.singaporemath.com

Dimensions Math® Teacher's Guide 6B
ISBN 978-1-947226-45-6

First published 2018
Reprinted 2019, 2020, 2021, 2022

Copyright © 2017 by Singapore Math Inc.

All rights reserved. This book or any portion thereof may not
be reproduced or used in any manner whatsoever without
the express written permission of the publisher.

Printed in China

Acknowledgments

Copy editing by the Singapore Math Inc. team.

How to Use This Teacher's Guide

Dimensions Math® 6 is based on the Singapore Mathematics Framework.

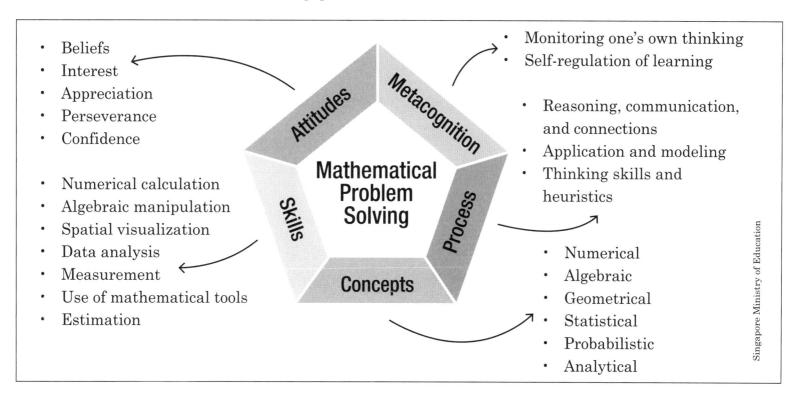

Dimensions Math® curriculum helps students become effective problem solvers by developing these five interrelated components. Activity-based learning, direct instruction, and teacher-led inquiry all deepen problem solving skills.

This teacher's guide offers lesson plans and helpful notes about mathematical ideas, possible strategies and errors, and practical tips to foster student understanding.

Each lesson typically follows a five-stage plan, although some lessons vary from this format slightly.

1. Introduction
   Introduce the concept.

2. Development
   Develop concepts, procedures, and skills through activities, problems, and discussion.

3. Application
   Apply concepts, skills, and procedures through real-life and mathematical problems or activities.

4. Extension
   Extend concepts, skills, and procedures through more complex problem situations.

5. Conclusion
   Summarize the main points of the lesson.

Most lessons are designed to be taught in a single 50-60 minute period. However, they can be adjusted to suit students' pace and class schedules.

The beginning of each chapter includes a section to help teachers understand important mathematical content.

The practice lessons do not follow the same format as other lessons. These lessons are the "exercises" in the textbook and are meant to consolidate and extend learning from previous lessons.

At the end of each chapter there are additional sections.

- In a Nutshell
  Review and summarize important mathematical ideas from the chapter.

- Write In Your Journal
  Encourage students to reflect on their learning and incorporate writing into the learning process.

- Extend Your Learning Curve
  Investigate interesting problems and engage in independent research.

- Problem Solving Corner
  Apply the Singapore model drawing method to solve complex problems related to the content of the chapter.

Calculators

Generally, problems in the Dimensions Math® series are written so that calculators are not necessary. However, calculators can be used at the teacher's discretion, especially for more complex problems. A calculator is not a substitute for basic facts knowledge, and struggling students should be given additional help to achieve fluency.

Use of Workbook

The workbook can be used in various ways, depending on the philosophy of the school and goals of the teachers and students. Teachers may consider these alternative uses at their discretion.

- Problems from the workbook can be assigned for homework.
- Students who finish problems early can do problems from the workbook.
- Problems from the workbook can be used in lieu of problems in the textbook.
- Additional practice lessons can be taught using the workbook.

Use of Notebooks

Students should use notebooks to record solution methods, important mathematical ideas that emerge from the lesson, and their reflections on their learning. Well-organized notebooks help students recall prior knowledge and track their own progress.

Notes

Contents

Chapter 8
Algebraic Expressions

Teaching Notes	1
Lesson 1	5
Lesson 2	12
Lesson 3	17
Lesson 4	21
Lesson 5	23
Lesson 6	28
Lesson 7	30

Chapter 9
Equations and Inequalities

Teaching Notes	33
Lesson 1	38
Lesson 2	42
Lesson 3	50
Lesson 4	54
Lesson 5	58
Lesson 6	60
Lesson 7	63
Lesson 8	65
Lesson 9	67

Chapter 10
Coordinates and Graphs

Teaching Notes	73
Lesson 1	79
Lesson 2	87
Lesson 3	90
Lesson 4	94
Lesson 5	103
Lesson 6	108
Lesson 7	111
Lesson 8	113
Lesson 9	119
Lesson 10	126
Lesson 11	132

Chapter 11
Area of Plane Figures

Teaching Notes	135
Lesson 1	140
Lesson 2	146
Lesson 3	151
Lesson 4	157
Lesson 5	161
Lesson 6	167
Lesson 7	171
Lesson 8	177
Lesson 9	185
Lesson 10	189

Chapter 12
Volume and Surface Area of Solids

Teaching Notes	191
Lesson 1	195
Lesson 2	201
Lesson 3	206
Lesson 4	211
Lesson 5	216
Lesson 6	221
Lesson 7	226
Lesson 8	234
Lesson 9	237
Lesson 10	239

Chapter 13
Displaying and Comparing Data

Teaching Notes	245
Lesson 1	249
Lesson 2	253
Lesson 3	257
Lesson 4	260
Lesson 5	265
Lesson 6	271
Lesson 7	279
Lesson 8	283
Lesson 9	286
Lesson 10	288
Lesson 11	292
Lesson 12	296
Lesson 13	303
Lesson 14	307

Notes

Chapter 8: Algebraic Expressions

Lesson	Objectives	Class Periods	Textbook & Workbook	Teacher's Guide Page	Additional Materials Needed
1	• Write algebraic expressions in which letters stand for numbers. • Write algebraic expressions to represent real world situations.	1	TB: 2–7 WB: 1–5	5	
2	• Evaluate algebraic expressions by substituting numbers for variables.	1	TB: 7–12 WB: 6–8	12	Toothpicks (or straws)
3	• Write algebraic expressions to solve real world problems. • Express one quantity in terms of another quantity.	1	TB: 13–16 WB: 9–13	17	Paper strips
4	• Consolidate and extend the material covered thus far.	1	TB: 17–18	21	Index cards
5	• Simplify algebraic expressions with up to three terms by combining like terms. • Determine whether two expressions are equivalent by simplifying the expressions. • Use the distributive property to write equivalent expressions.	2	TB: 20–24 WB: 14–18	23	
6	• Consolidate and extend the material covered thus far.	1	TB: 24–25	28	
7	• Summarize and reflect on important ideas learned in this chapter. • Apply and extend understanding by investigating a non-routine problem.	1	TB: 26–27	30	

©2017 Singapore Math Inc. Dimensions Math® Teacher's Guide 6B

Chapter 8: Algebraic Expressions

Algebra is often called "generalized arithmetic." The same properties of operations that hold true for arithmetic hold true for algebra. That is why a student's best preparation for algebra is a strong understanding of arithmetic. This foundation was laid in **Dimensions Math® 6A** Chapters 1 – 3, especially Chapter 1.

Students have already seen variables before in formulas such as the area of rectangles ($l \times w$), area of squares ($s \times s = s^2$), and perimeter of rectangles ($2l + 2w$) or $2(l + w)$.

They have also seen variables when they generalized the properties of operations in Chapter 1. These properties are very important for students to understand because they also hold true for algebra.

Properties of Addition

- **Commutative Property of Addition**
  $a + b = b + a$
  For example: $2 + 3 = 3 + 2$

- **Associative Property of Addition**
  $(a + b) + c = a + (b + c)$
  For example: $(2 + 3) + 4 = 2 + (3 + 4)$

Properties of Multiplication

- **Commutative Property of Multiplication**
  $a \times b = b \times a$
  For example: $5 \times 4 = 4 \times 5$

- **Identity Property of Multiplication**
  $a \times 1 = a$, where $a \neq 0$
  For example: $5 \times 1 = 5$

- **Zero Property of Multiplication**
  $a \times 0 = 0$
  For example: $7 \times 0 = 0$

- **Associative Property of Multiplication**
  $(a \times b) \times c = a \times (b \times c)$
  For example: $(2 \times 3) \times 4 = 2 \times (3 \times 4)$

- **Distributive Property of Multiplication**
  $a \times (b + c) = a \times b + a \times c$
  For example: $5 \times (3 + 2) = 5 \times 3 + 5 \times 2$
  $(6 - 4) \times 8 = 6 \times 8 - 4 \times 8$

Properties of Division

- **Identity Property of Division**
  $a \div 1 = a$ where $a \neq 0$
  For example: $5 \div 1 = 5$

- **Zero Property of Division**
  $0 \div a = 0$, where $a \neq 0$
  For example: $0 \div 5 = 0$

- We can divide the sum of, or the difference between, two numbers in parentheses by dividing each number in the parentheses.
  For example: $(8 + 4) \div 2 = 8 \div 2 + 4 \div 2$
  $(20 - 10) \div 5 = 20 \div 5 - 10 \div 5$

- When we multiply or divide both the dividend and the divisor by the same number, the quotient remains the same.
  For example: $24 \div 8 = (24 \div 4) \div (8 \div 4)$
  $6 \div 2 = (6 \times 3) \div (2 \times 3)$

Chapter 8: Algebraic Expressions

The teacher should help students relate what they learned previously about numerical expressions to help them understand algebraic expressions. An important difference between numerical expressions and algebraic expressions is that numerical expressions always have a fixed value. In algebraic expressions, variables can take on different values.

When we replace the variable in an algebraic expression with a numerical value, we **evaluate** the expression. This process is called **substitution**. For example, to evaluate the expression $x + 5$ when $x = 2$, we substitute 2 for x to get $2 + 5 = 7$.

Algebraic expressions are helpful to model real life situations in which quantities can change. For example, the total pay of an employee who earns $20 per hour can be expressed as $(20h)$. This expression can be used to determine the total pay given different numbers of hours the employee works. If she works 40 hours, the total pay will be (20×40). If the number of hours she works changes, we can still use the same expression to find her total pay.

We can use expressions to compare two quantities. When we do this, we say we are expressing one quantity **in terms of** the other quantity.

For example, if Sarah is 3 years older than her brother Timmy, we can represent Sarah's age in terms of Timmy's age as $(t + 3)$ years. This relationship will always hold true (assuming they share the same birthday). We are comparing Sarah's age to her brother's age, so her brother's age is the base quantity or quantity we are comparing to.

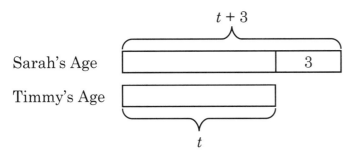

We can also compare Timmy's age to his sister's age. Timmy will always be 3 years younger than his sister.

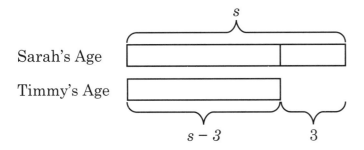

In this case, we are considering Sarah's age as the base quantity. We can express Timmy's age in terms of Sarah's age as $s - 3$.

Chapter 8: Algebraic Expressions

In algebraic expressions we write multiplication terms using a coefficient and a variable. $5 \times n$, for example, is written as $5n$, 5 being the coefficient and n the variable. The coefficient tells us how many times the variable is added repeatedly. It can also be interpreted as the number that is being added repeatedly. Thus $5n$ can mean $n + n + n + n + n$ or 5 added repeatedly n times. If the variable has no coefficient, we can think of its coefficient as 1. Thus both $1 \times a$ and $a \times 1$ are written as a.

Division expressions are generally written as fractions. The expression $a \div 2$ would normally be written as $\frac{a}{2}$, the expression $(y - 2) \div 3$ as $\frac{y-2}{3}$, and the expression $4 \div (a + 3)$ as $\frac{4}{a+3}$.

We can simplify algebraic expressions by finding a simpler equivalent expression. Equivalent expressions are expressions that have the same value. We simplify expressions involving addition and subtraction by grouping and combining like terms. Like terms are terms where the variable is the same and to the same power. $5x$ and $3x$ are like terms. $5x$ and $3y$ are not like terms because they do not have the same variable. $5x$ and $3x^2$ are not like terms because they are not to the same power.

We can simplify expressions like $5x + 7 + 3x - 3$ by grouping and combining the variable terms and the constant (numerical) terms.

$$5x + 7 + 3x - 3 = (5x + 3x) + (7 - 3) = 8x + 4$$

The expression $5x + 7 + 3x - 3$ and the resulting simplified expression $8x + 4$ are equivalent. If you substitute the same value for x on each side of the equation (e.g., $x = 2$), the values will be equal.

$$5x + 7 + 3x - 3 = 8x + 4$$
$$5(2) + 7 + 3(2) - 3 = 8(2) + 4$$
$$20 = 20$$

We can simplify expressions involving multiplication with parentheses by using the distributive property of multiplication to remove the parentheses.

$$4\left(2y + \frac{1}{2}\right) = 4 \times 2y + 4 \times \frac{1}{2} = 8y + 2$$

The expression $4\left(2y + \frac{1}{2}\right)$ and the resulting simplified expression $8y + 2$ are equivalent. The expression is simplified when like terms are combined, fractions are simplified, and brackets are removed.

In **Dimensions Math® 6**, students will not be simplifying expressions involving exponents. These ideas will be learned in **Dimensions Math® 7**.

Lesson 1

Objectives:

- Write algebraic expressions in which letters stand for numbers.
- Write algebraic expressions to represent real world situations.

1. Introduction

Read and discuss the Chapter Opener.

State the learning goals of the chapter.

Student Textbook page 1

Discuss pages 2 – 3 and ask students to draw the bar models on page 2 to model the algebraic expressions.

Have students talk about DISCUSS with a partner or in groups.

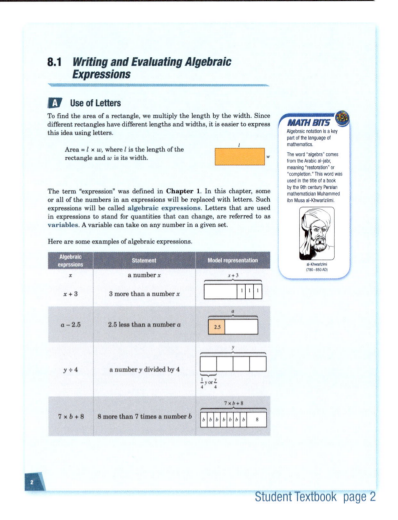

Student Textbook page 2

Notes:

- When a variable does not have a coefficient, we consider the coefficient as 1 although it is not written. We write $1x$ as x.
- A single term like x or $5x$ is also considered an algebraic expression.
- Dividing a number by 4 is the same as finding $\frac{1}{4}$ of a number, thus $y \div 4 = \frac{y}{4} = \frac{1}{4} \times y = \frac{1}{4}y$.
- When a variable and a number are multiplied, we always write the number before the variable. Thus, $7b$ can mean $b \times 7$ or $7 \times b$.
- We can write $\frac{1}{7} \times n$ as $\left(\frac{1}{7}\right)n$, but it is usually written without the parentheses as $\frac{1}{7}n$ or $\frac{n}{7}$.

2. Development

Have students study Examples 1 – 4 and do Try It! 1 – 4.

Example 1:
- With addition expressions, the order of the terms can be switched based on the commutative property of addition so **(c)** can also be expressed as $n + 5$ although $5 + n$ better models the situation "n more than 5."
- The commutative property does not apply to subtraction, so **(d)** cannot be expressed as $y - 1$.

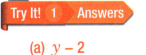

(a) $y - 2$ (b) $2.8 - p$

(c) $12 + x$ (d) $w + \frac{3}{4}$

Student Textbook page 3

Example 2:

- For Example 2 (c) and Try It! 2 (c), students may mistakenly write 2y or 4r. Refer them to the first REMARKS on page 4.

- For Try It! 2 (d), $\frac{3q}{4}$ is also acceptable, $\frac{3}{4q}$ is not. See REMARKS on page 5.

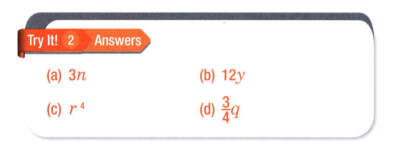

Try It! 2   Answers

(a) $3n$ (b) $12y$

(c) r^4 (d) $\frac{3}{4}q$

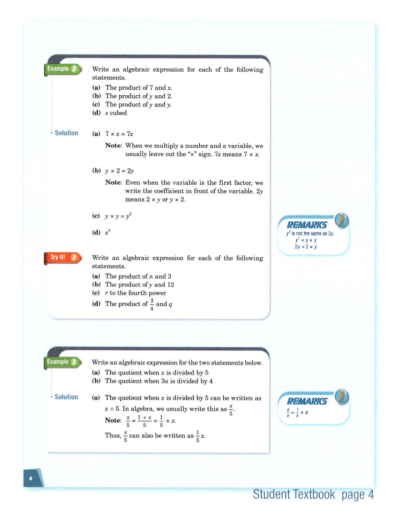

Student Textbook page 4

Example 4:

- Bring attention to the notes and models written in blue, and help students see the connection between the algebraic expressions and the bar models.

Try It! 3 Answers

(a) $\dfrac{y}{7} = \dfrac{1}{7}y$ (b) $\dfrac{b}{2} = \dfrac{1}{2}b$

(c) $\dfrac{4x}{9} = \dfrac{4}{9}x$ (d) $\dfrac{3w}{5} = \dfrac{3}{5}w$

Try It! 4 Answers

(a) $6b + \dfrac{2}{3}$ (b) $\dfrac{1}{7}y - 4.1$

(c) $12 - \dfrac{3}{8}p$ (d) $10 + \dfrac{6n}{11}$

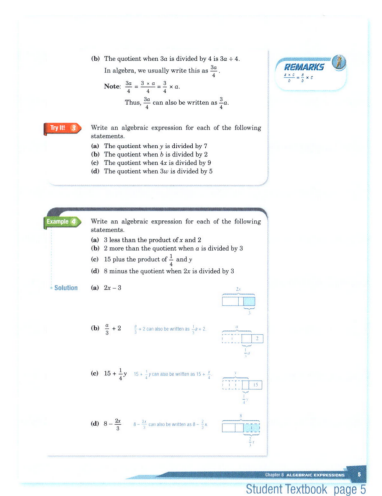

Student Textbook page 5

3. Application

Have students study Examples 5 – 7 and do Try It! 5 – 7.

Example 5:
- Ask students, "What if Lina ran the race in 12 minutes? Did she run faster or slower than Kelly? Whose time will be greater, Kelly's or Lina's?" Help them see that since Lina ran slower, her time would be greater. Thus, the correct expression is $q - 15$, not $15 - q$.
- The unit (minutes) applies to both q and 15, so we use parentheses to write $(q - 15)$ minutes, not $q - 15$ minutes (see REMARKS).

Example 6:
- These problems have two different variables. If students are confused, have them substitute numbers for the variable cents (e.g., if an apple costs 20¢ and a cookie costs 50¢, the expression is $10 \times 20 + 1 \times 50$ cents). Then put the variables back: $(10 \times a + 1 \times c = 10a + c)$ cents.

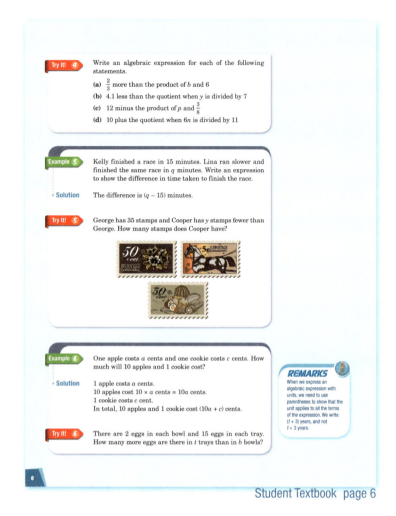

Student Textbook page 6

Try It! 5 Answer

$(35 - y)$ stamps

Try It! 6 Answer

$(15t - 2b)$ eggs

Example 7:

- Students can write the expression using the ÷ symbol before writing it in fractional form. The amount Sally had left after eating 3 candies was $(y - 3)$ candies. After dividing them among 4 friends, the number of candies for each person is $(y - 3) \div 4 = \frac{y - 3}{4}$.

Try It! 7:

- The number of buttons Karen had left was $(k - 6)$ buttons. After splitting them among her friends, the number of buttons became $(k - 6) \div 5 = \frac{k - 6}{5}$.

Try It! 7 Answer

$\frac{k - 6}{5}$ buttons

Student Textbook page 7

4. Conclusion

Summarize the important points of the lesson.

- We can write algebraic expressions with variables to model quantities and real life situations.
- We can replace the variable in the expression with any number to model the situation given different conditions. For example, the total pay of an employee who earns $20 per hour can be expressed as $(20h)$. This expression can be used to determine the total pay given different numbers of hours the employee works.

★ **Workbook: Page 1**

Lesson 2

Objective:

• Evaluate algebraic expressions by substituting numbers for variables.

1. Introduction

Discuss the idea of substitution using the example of evaluating $x + 2$ when $x = 3$ on page 7. Ask students to substitute other numbers for x.

Note:

• When we evaluate an expression, we are finding the value of the expression when the numerical value of the variable is given.

Ask students to come up with other similar examples on their own using expressions involving the four operations (e.g., evaluate $\frac{y - 3}{4}$ (when $y = 15$).

2. Development

Give students toothpicks (or have them draw them) and complete Class Activity 1 with a partner or group.

Have students share their answers and conclusions from the activity.

Summarize the results of the class activity using page 9 and discuss what the character is saying.

Answers for Class Activity 1

2. $1 + \underline{3}$
3. $1 + 3 \times \underline{2}$
4. $1 + 3 \times \underline{3}$
5. (a) $1 + 3 \times 4$, $1 + 3 \times 5$
   (b) $1 + 3 \times 22 = 67$
   (c) $1 + 3 \times n = 1 + 3n$
6. (a) $1 + 3 \times 7 = 22$ (b) $1 + 3 \times 13 = 40$
   (c) $1 + 3 \times 27 = 82$ (d) $1 + 3 \times 311 = 934$

---

2. Now place 3 more toothpicks next to it to form a square as shown.

Complete the expression below to represent the first toothpick, and then the toothpicks added to create 1 square.

$$1 + \underline{\quad}$$

First toothpick   Added toothpicks

3. Add 3 more toothpicks to create another square as shown.

Complete the expression below to represent the total number of toothpicks for 2 squares.

$$1 + 3 \times \underline{\quad}$$

First toothpick   Added toothpicks

4. Add 3 more toothpicks to create another square.

Complete the expression below to represent the total number of toothpicks for 3 squares.

$$1 + 3 \times \underline{\quad}$$

First toothpick   Added toothpicks

Student Textbook page 8

Number of Squares	Expression	Number of Toothpicks
1	$1 + 3 \times 1$	4
2	$1 + 3 \times 2$	7
3	$1 + 3 \times 3$	10
4	$1 + 3 \times 4$	13
5	$1 + 3 \times 5$	16
22	$1 + 3 \times 22$	67

©2017 Singapore Math Inc. Dimensions Math® Teacher's Guide 6

Notes:

- Point out that we can generalize the pattern. Using the expression $1 + 3n$, the expression for 1 square can be written as $1 + 3 \times 1$ instead of $1 + 3$.
- Point out that the expression $1 + 3n$ will work for any number of squares.
- Ask students what the expression would be for 0 squares. $1 + 3n = 1 + 3(0) = 1$ (i.e., the first toothpick in #1.)

3. Application

Have students study Examples 8 – 10 and do Try It! 8 – 10.

Example 8:

- For Example (c) and Try It! (c), students may mistakenly write 3×5 or 2×6.
- For Example (d), students could alternatively write $\frac{3}{5} \times 5$. For Try It! (d), they could write $\frac{2 \times 6}{3}$ or $\frac{2}{3} \times 6$.

Try It! 8 Answers

(a) $7 \times 6 = 42$

(b) $10 - 6 = 4$

(c) $6 \times 6 = 36$

(d) $\frac{2 \times 6}{3} = 4$

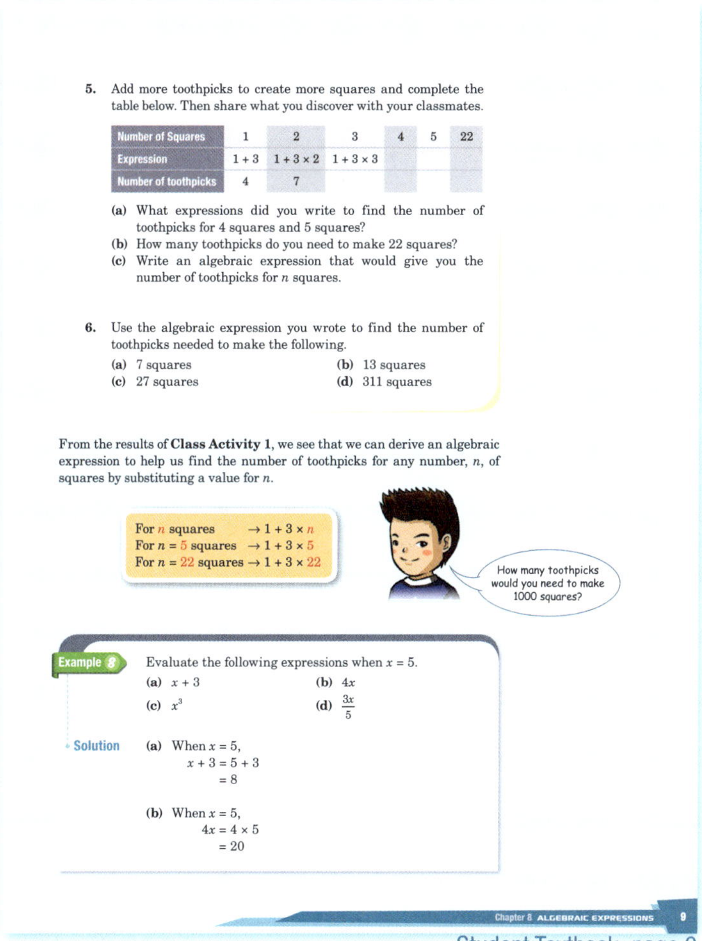

5. Add more toothpicks to create more squares and complete the table below. Then share what you discover with your classmates.

Number of Squares	1	2	3	4	5	22
Expression	$1 + 3$	$1 + 3 \times 2$	$1 + 3 \times 3$			
Number of toothpicks	4	7				

(a) What expressions did you write to find the number of toothpicks for 4 squares and 5 squares?
(b) How many toothpicks do you need to make 22 squares?
(c) Write an algebraic expression that would give you the number of toothpicks for n squares.

6. Use the algebraic expression you wrote to find the number of toothpicks needed to make the following.
(a) 7 squares
(b) 13 squares
(c) 27 squares
(d) 311 squares

From the results of **Class Activity 1**, we see that we can derive an algebraic expression to help us find the number of toothpicks for any number, n, of squares by substituting a value for n.

For n squares $\rightarrow 1 + 3 \times n$
For $n = 5$ squares $\rightarrow 1 + 3 \times 5$
For $n = 22$ squares $\rightarrow 1 + 3 \times 22$

Example 8 Evaluate the following expressions when $x = 5$.

(a) $x + 3$
(b) $4x$
(c) x^3
(d) $\frac{3x}{5}$

Solution

(a) When $x = 5$,
$x + 3 = 5 + 3$
$= 8$

(b) When $x = 5$,
$4x = 4 \times 5$
$= 20$

Chapter 8 ALGEBRAIC EXPRESSIONS 9

Student Textbook page 9

Example 9:

- For Example 9 (a), students may alternatively write $\frac{3 \times 8}{4}$. They could use the same method in Try It! (a) and (c).
- For Example 9 (b), many students already know that $0.5 = \frac{1}{2}$, so they may do $\frac{1}{2} \times 8$. In Try It! 9 (b), they may do the same ($0.75 = \frac{3}{4}$).

Try It! 9 Answers

(a) $\frac{1}{2} \times 12 = 6$ (b) $0.75 \times 12 = 9$

(c) $\frac{7}{6} \times 12 = 14$ (d) $1.4 \times 12 = 16.8$

Try It! 10 Answers

(a) $25.7 - 3 \times 8 = 25.7 - 24 = 1.7$

(b) $9 + \frac{8}{8} = 9 + 1 = 10$

(c) $3 \times 8 + \frac{1}{2} = 24 + \frac{1}{2} = 24\frac{1}{2}$

(d) $10 - \frac{3 \times 8}{4} = 10 - \frac{24}{4} = 4$

(e) $\frac{1}{3} \times 8 - \frac{2}{3} = \frac{8}{3} - \frac{2}{3} = \frac{6}{3} = 2$

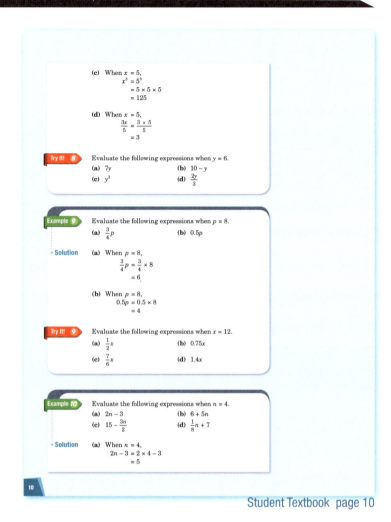

Student Textbook page 10

4. Extension

Have students study Examples 11 – 12 and do Try It! 11 – 12.

In these problems, the value being substituted for the variable is a fraction. Look for mechanical errors and provide additional support for students who are struggling with fraction calculations.

Example 11:

- Make sure students are following the correct order of operations. For Example 11 (b), ask students how we would evaluate the expression if we change the order of the terms to $3 + \frac{1}{2}x$.

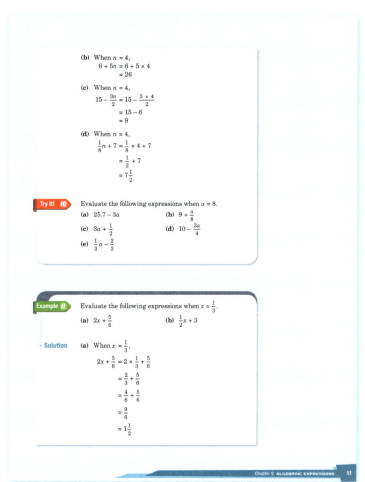

Student Textbook page 11

Try It! 11 Answers

(a) $\frac{2}{5} - \frac{1}{5} = \frac{1}{5}$

(b) $10 \times \frac{2}{5} - 1 = 4 - 1 = 3$

(c) $\frac{15}{2} \times \frac{2}{5} + 3 = 3 + 3 = 6$

(d) $\frac{1}{3} \times \frac{2}{5} + \frac{1}{5} = \frac{2}{15} + \frac{1}{5} = \frac{2}{15} + \frac{3}{15} = \frac{5}{15} = \frac{1}{3}$

Example 12:
- Method 1 generally involves fewer steps and is probably the easiest method for most students.
- Method 2 changes the fractional expression to a division expression to eliminate the fractional term.
- Encourage students to carefully write out the steps to transform the equations.

5. Conclusion

Summarize the important points of the lesson.
- We can use algebraic expressions to model quantities that change (e.g., the total number of matchsticks when the number of squares changes).
- When we evaluate an expression, we are finding the value of the expression for a given numerical value of the variable.
- Follow the order of operations and write each step clearly and carefully.

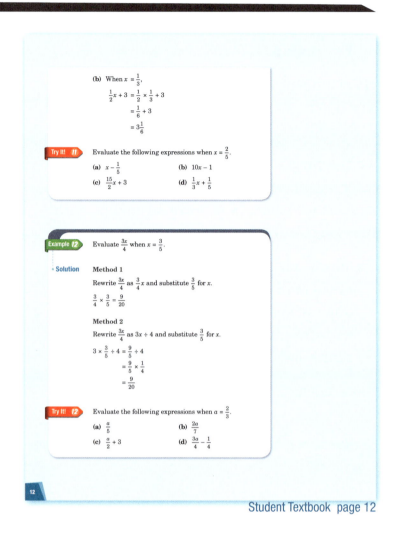

Student Textbook page 12

Try It! 12 Answers (Method 1)

(a) $\frac{a}{5} = \frac{1}{5}a, \frac{1}{5} \times \frac{2}{3} = \frac{2}{15}$

(b) $\frac{2a}{7} = \frac{2}{7}a, \frac{2}{7} \times \frac{2}{3} = \frac{4}{21}$

(c) $\frac{a}{2} + 3 = \frac{1}{2}a + 3, \frac{1}{2} \times \frac{2}{3} + 3 = \frac{2}{6} + 3 = \frac{1}{3} + 3 = 3\frac{1}{3}$

(d) $\frac{3a}{4} - \frac{1}{4} = \frac{3}{4}a - \frac{1}{4}, \frac{3}{4} \times \frac{2}{3} - \frac{1}{4} = \frac{6}{12} - \frac{1}{4} = \frac{1}{4}$

Try It! 12 Answers (Method 2)

(a) $\frac{a}{5} = a \div 5, \frac{2}{3} \div 5 = \frac{2}{3} \times \frac{1}{5} = \frac{2}{15}$

(b) $\frac{2a}{7} = 2a \div 7, 2 \times \frac{2}{3} \div 7 = \frac{4}{3} \div 7 = \frac{4}{3} \times \frac{1}{7} = \frac{4}{21}$

(c) $\frac{a}{2} + 3 = a \div 2 + 3, \frac{2}{3} \div 2 + 3 = \frac{2}{3} \times \frac{1}{2} + 3 = \frac{1}{3} + 3 = 3\frac{1}{3}$

(d) $\frac{3a}{4} - \frac{1}{4} = 3a \div 4 - \frac{1}{4}, (3 \times \frac{2}{3}) \div 4 - \frac{1}{4} = 2 \div 4 - \frac{1}{4} = \frac{2}{4} - \frac{1}{4} = \frac{1}{4}$

★ **Workbook: Page 6**

Lesson 3

Objectives:

- Write algebraic expressions to solve real world problems.
- Express one quantity in terms of another quantity.

1. Introduction

Give students 2 paper strips, one slightly longer than the other. Tell them that these strips represent the ages of Sarah and her brother Timmy. Sarah is 3 years older than Timmy.

Have students fold Sarah's paper to make it the same length as Timmy's, and write 3 on the extra part.

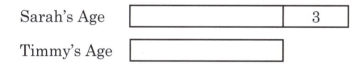

Ask the students what possible ages Sarah and Timmy could be.

2. Development

Have students solve and discuss Example 13. (They can cover the answer.) Then, discuss the solution given in the textbook.

Notes:
- Expressing Sarah's age in terms of Timmy's age means that we are considering Timmy's age is the base (quantity we are comparing to). Thus, Sarah's age becomes $(t + 3)$ years.
- We could also express Timmy's age in terms of Sarah's age by making Sarah's age the base. Thus, Timmy's age would be $(s - 3)$ years.
- Since we can express the quantities in different ways, it is important to define the variable precisely.

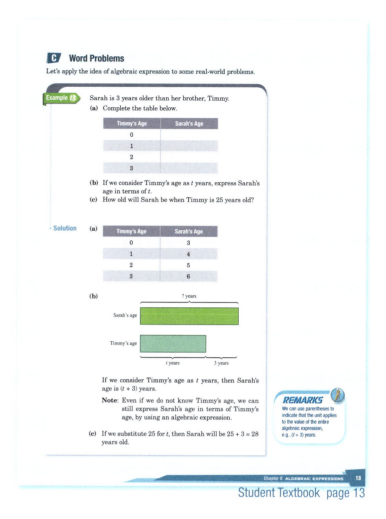

Have students read Try It! 13 and draw a bar model to represent the temperatures. Then, have them solve the problem, and share and discuss their solutions.

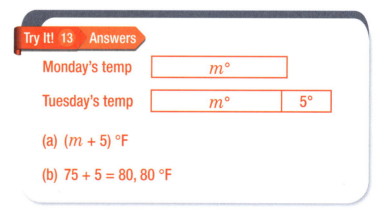

3. **Application**

Have students study Examples 14 – 15 and do Try It! 14 – 15.

Example 14:
- In this case, the larger quantity (blue ribbon) is considered the base, so the smaller quantity becomes $b - 5$ (cm).
- If students are having difficulty with Try It! 14, ask them to draw a bar model.

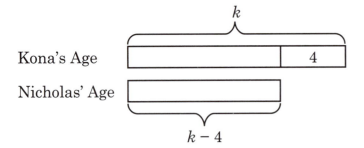

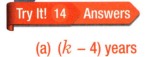

(a) $(k - 4)$ years

(b) $17 - 4 = 13$, 13 years old

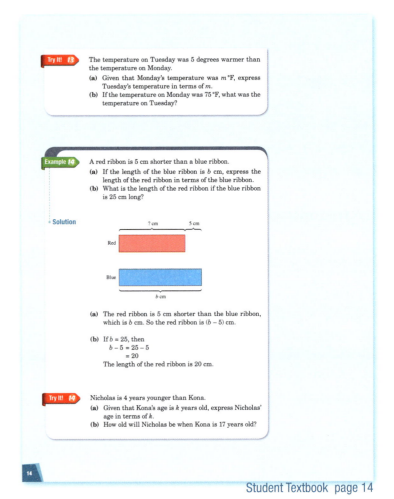

Student Textbook page 14

Example 15:

- In these problems, students apply a familiar formula, $l \times w$, to find the area of rectangles. Since there are many different dimensions that a rectangle can have, we can generalize the area of all rectangles using this formula. Although they have seen this formula since Grade 3, they may not realize that they were using an algebraic expression.
- Give students other possible values for the width of the rectangle in Example 15 and the length of the rectangle in Try It! 15. Ask them to find the areas.

Try It! 15   Answers

(a) Area of paper rectangle = $l \times 2 = 2l$

(b) $2l$ in²

(c) When $l = 12$, $2l = 2 \times 12 = 24$; 24 in²

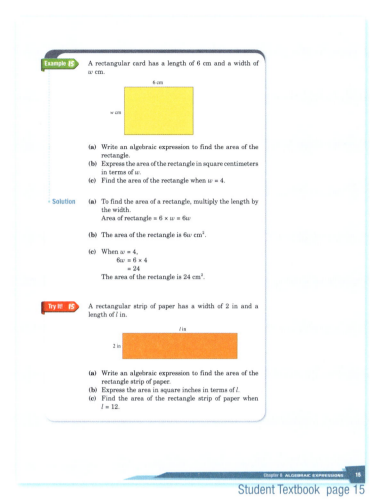

Student Textbook page 15

4. Extension

Have students study Example 16 and then do Try It! 16.

Example 16:

- In this case, the smaller quantity (girls) is considered the base, so the larger quantity (boys) becomes $3 \times g = 3g$. To help students understand this, have them redraw the model by drawing a bar for the girls (1 unit) first, then drawing a bar 3 times as long (3 units) for the boys.

- As an extension, ask students to express the number of girls in terms of the number of boys, if the number of boys is b $\left(\frac{1}{3} \times b = \frac{1}{3}b \text{ or } b \div 3 = \frac{b}{3}\right)$.

Try It! 16:

- Ask students to draw a bar model for Try It! 16 before writing the expression. Ask them to substitute different numbers for t.

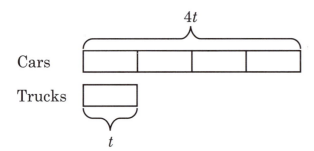

- Ask students to express the number of trucks in terms of the number of cars, if the number of cars is c $\left(\frac{1}{4} \times c = \frac{1}{4}c \text{ or } c \div 4 = \frac{c}{4}\right)$.

Try It! 16   Answers

(a) There are $4 \times t = 4t$ cars.

(b) If $t = 55$, $4t = 4 \times 55 = 220$; 220 cars

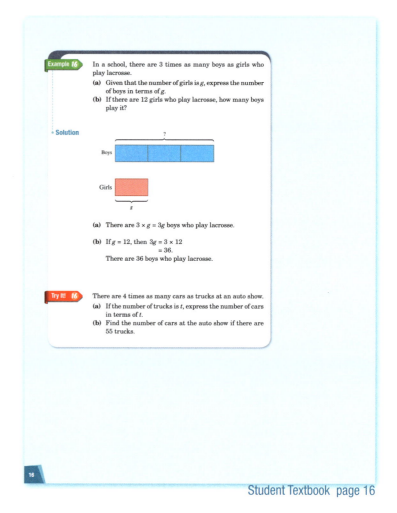

Student Textbook page 16

5. Conclusion

Summarize the important points of the lesson.

- We can apply algebraic expressions to real world problems. For example, the expression $t + 3$ on page 13 will give us Sarah's age for any age of Timmy.
- When we are comparing two quantities, we consider one quantity as the base. We can then express the other quantity in terms of that quantity.

★ **Workbook: Page 9**

Lesson 4

> **Objective:**
> - Consolidate and extend the material covered thus far.

Have students work together with a partner or in groups. Students should try to solve the problems by themselves first, then compare solutions with their partner or group. If they are confused, they can discuss together.

Observe students carefully as they work on the problems. Give help as needed individually or in small groups.

 BASIC PRACTICE

1. (a) $y + 7$ (b) $x - 5$
   (c) $9.1 + w$ (d) $\frac{3}{4} - p$

2. (a) $17n$ (b) $8p$
   (c) n^6 (d) $\frac{5}{8}q$

3. (a) $\frac{3x}{5} = \frac{3}{5}x$ (b) $\frac{x}{3} = \frac{1}{3}x$
   (c) $\frac{2y}{7} = \frac{2}{7}y$ (d) $\frac{a}{4} = \frac{1}{4}a$

4. (a) 10   (b) 15   (c) 9
   (d) 81   (e) 1   (f) $2\frac{1}{2}$

 FURTHER PRACTICE

5. (a) $12 + \frac{7}{10}x$ (b) $\frac{1}{3}x + \frac{3}{5}$
   (c) $6 - 0.4m$ (d) $\frac{n}{8} - 8.2$
   (e) $x^2 + 2$

6. (a) $30 - 2 \times 5 = 20$ (b) $5 + \frac{2}{3} = 5\frac{2}{3}$
   (c) $8 + \frac{5}{10} = 8\frac{1}{2}$ (d) $\frac{4 \times 5}{5} - 4 = 0$
   (e) $\frac{3}{4} \times 5 + \frac{1}{4} = 4$ (f) $0.7 \times 5 + 2.3 = 5.8$
   (g) $\frac{3}{10} \times 5 + 0.8 = 2.3$ (h) $5^2 - 5 = 20$

 **EXERCISE 8.1**

BASIC PRACTICE

1. Write an algebraic expression for each of the following statements.
   (a) 7 more than y
   (b) 5 less than x
   (c) w more than 9.1
   (d) p less than $\frac{3}{4}$

2. Write an algebraic expression for each of the following statements.
   (a) The product of n and 17
   (b) The product of 8 and p
   (c) n to the sixth power
   (d) The product of q and $\frac{5}{8}$

3. Write an algebraic expression for each of the following statements.
   (a) The quotient when $3x$ is divided by 5
   (b) The quotient when x is divided by 3
   (c) The quotient when $2y$ is divided by 7
   (d) The quotient when a is divided by 4

4. Evaluate the following expressions when $a = 3$.
   (a) $a + 7$ (b) $5a$
   (c) $12 - a$ (d) a^4
   (e) $\frac{a}{3}$ (f) $\frac{5}{6}a$

FURTHER PRACTICE

5. Write an algebraic expression for each of the following statements.
   (a) 12 plus the quotient of $7x$ by 10
   (b) $\frac{3}{5}$ more than the product of x and $\frac{1}{3}$
   (c) 6 minus the product of m and 0.4
   (d) 8.2 less than the quotient of n by 8
   (e) 2 more than x squared

6. Evaluate the following expressions when $x = 5$.
   (a) $30 - 2x$
   (b) $x + \frac{2}{3}$
   (c) $8 + \frac{x}{10}$
   (d) $\frac{4x}{5} - 4$
   (e) $\frac{3}{4}x + \frac{1}{4}$
   (f) $0.7x + 2.3$
   (g) $\frac{3}{10}x + 0.8$
   (h) $x^2 - 5$

7. Evaluate the following expressions when $x = \frac{3}{4}$.
   (a) $4x + 3$ (b) $x - \frac{1}{2}$
   (c) $7x - 3$ (d) $0.5 + 12x$
   (e) $\frac{2}{3}x - \frac{1}{4}$ (f) $\frac{4x}{5}$
   (g) $\frac{x}{3} + 1$ (h) $\frac{8x}{3} - 0.5$

 **MATH @ WORK**

8. Nicole and Grace share a birth date, but Grace is 5 years older.
   (a) If Nicole is 15 years old, how old is Grace?
   (b) If Nicole is p years old, how old is Grace?
   (c) If Grace is q years old, how old is Nicole?

Student Textbook page 17

7. (a) $4 \times \frac{3}{4} + 3 = 6$ (b) $\frac{3}{4} - \frac{1}{2} = \frac{1}{4}$
   (c) $7 \times \frac{3}{4} - 3 = 2\frac{1}{4}$ (d) $0.5 + 12 \times \frac{3}{4} = 9.5$
   (e) $\frac{2}{3} \times \frac{3}{4} - \frac{1}{4} = \frac{1}{4}$ (f) $\frac{4}{5} \times \frac{3}{4} = \frac{3}{5}$
   (g) $\frac{1}{3} \times \frac{3}{4} + 1 = 1\frac{1}{4}$ (h) $\frac{8}{3} \times \frac{3}{4} - 0.5 = 1.5$

 MATH @ WORK

8. (a) 20 years old   (b) $(p + 5)$ years old
   (c) $(q - 5)$ years old

9.
Number of Bags	Total Number of Oranges
1	5
2	10
3	15
4	20
5	25

9. (continued)

   (a) $15 \times 5 = 75$; 75 oranges

   (b) $k \div 5 = \frac{k}{5}$ bags

   Assuming k is a multiple of 5, it will be $k \div 5 = \frac{k}{5}$ bags. If it is not a multiple of 5, he can only pack a whole number of bags. E.g., if there are 16 oranges, $16 \div 5 = 5$ with remainder of 1, which means he has enough oranges to pack 5 bags and there will be 1 orange left over.

   (c) (i) $400 \div 5 = 80$; 80 bags

   (ii) $403 \div 5 = 80$ R 3; 3 oranges will be left over which is not enough for another bag, so it is still 80 bags.

   (iii) Any fractional amount or amount left over would not be enough for another bag (e.g., if there are 26 oranges $\frac{26}{5} = 5\frac{1}{5}$ or $26 \div 5 = 5$ R 1 so it would be 5 bags, not 6).

10. (a) 9 cycles ⟶ 15

    1 cycle ⟶ $\frac{15}{9}$

    21 cycles ⟶ $\frac{15}{9} \times 21 = 35$

    (b) 15 ⟶ 9 cycles

    1 ⟶ $\frac{9}{15}$ cycles

    t ⟶ $\frac{9}{15} \times t = \frac{3}{5}t$ cycles

    (c) $\frac{9}{15} \times 55 = 33$ cycles

11.

Number of Hours	Amount Jamie Charges in Dollars
1	24
2	48
5	120
p	$24p$

The expression for p hours is $\$24p$

(a) $24 \times 2\frac{1}{2} = 60$; $60   (b) $\frac{108}{24} = 4\frac{1}{2}$; $4\frac{1}{2}$ hours

Student Textbook page 18

12. (a) No. The correct answer is $2^3 + 4 = 8 + 4 = 12$.

    (b) He probably multiplied x by 3 instead of taking x to the third power.

    $2 \times 3 + 4 = 10$

13. $5c = 4s$

By looking at the model, we can see that $c = \frac{4}{5}s$ so $5c = 4s$. (Students could also write $\frac{c}{4} = \frac{s}{5}$.)

22

Lesson 5

Objectives:

- Simplify algebraic expressions with up to three terms by combining like terms.
- Determine whether two expressions are equivalent by simplifying the expressions.
- Use the distributive property to write equivalent expressions.

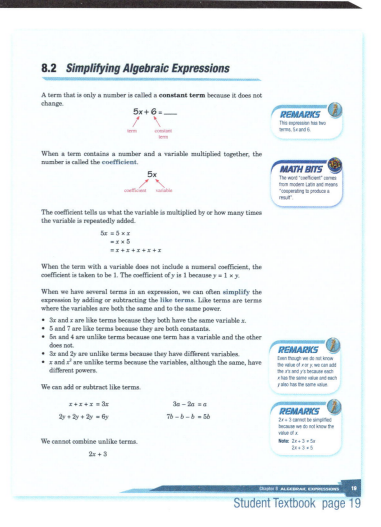
Student Textbook page 19

1. Introduction

Have students read and discuss page 19.

Notes:

- The value of a term with a variable can change depending on the value of the variable. A term without a variable always has the same value so it is called a **constant** term.
- Help students understand why we cannot combine a term with a variable and a constant term (see REMARKS at bottom of page 19).
- We can combine terms that have the same variable because the variable always has the same value (see REMARKS on bottom right of page 19). To see this more clearly, substitute numbers for the variables in the examples at the bottom of the page. For example, if $y = 3$:
$2y + 2y + 2y = (2 \times 3) + (2 \times 3) + (2 \times 3) = (2 + 2 + 2) \times 3 = 6 \times 3$.
Thus, $2y + 2y + 2y = (2 + 2 + 2) \times y = 6 \times y = 6y$.
Alternatively, $2y + 2y + 2y = (y + y) + (y + y) + (y + y) = 6 \times y = 6y$.

2. Development

Give pairs (or small groups) of students index cards to write the terms and play the game from Class Activity 2.

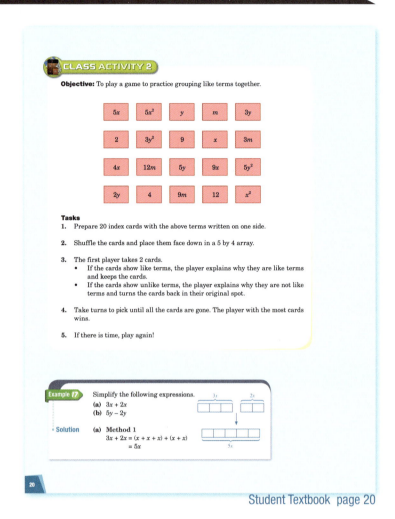

Student Textbook page 20

3. Application

Have students study Examples 17 – 20 and do Try It! 17 – 20.

Example 17:
- The original expression and the simplified expression are equivalent expressions (see note on page 21 after Try It! 17). To be equivalent expressions, the expressions have to remain equal for any value of x.
- Method 1 is the simplest method and is especially helpful for students who are struggling. Encourage students who are using this method to draw a bar model similar to the ones shown in the textbook.
- Method 2 is based on the distributive property of multiplication. In $3x + 2x$, the variable x is a common factor in each term. To see this more clearly, substitute a number for the variable:
- If $x = 5$, $3x + 2x = 3 \times 5 + 2 \times 5 = (3 + 2) \times 5$. Similarly for (b), y is a common factor of the terms in $5y - 2y$. If $y = 4$, then $5y - 2y = 5 \times 4 - 2 \times 4 = (5 - 2) \times 4$.

Example 18:
- These problems involve 3 terms. Remind students that you cannot combine a constant term with a term that has a variable.

DISCUSS
- $x - x$ and $0x$ are equivalent expressions but $0x = 0$. Thus, Peter did not simplify the expression completely. Similarly, $3p - 3p = 0p = 0$.

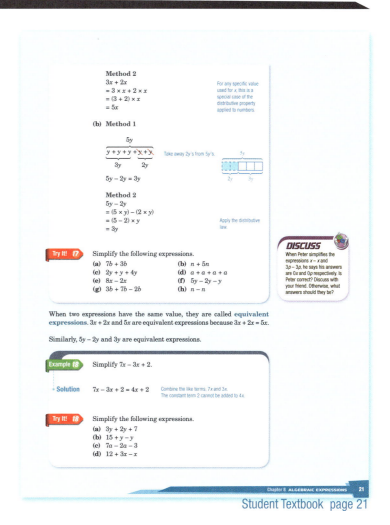

Student Textbook page 21

Try It! 17 Answers

(a) $10b$ (b) $6n$ (c) $7y$

(d) $4a$ (e) $6x$ (f) $2y$

(g) $8b$ (h) 0

Try It! 18 Answers

(a) $3y + 2y + 7 = 5y + 7$

(b) $15 + y - y = 15 + 0 = 15$

(c) $7a - 2a - 3 = 5a - 3$

(d) $12 + 3x - x = 12 + 2x$

Example 19:
- We can add in any order based on the commutative property of addition, thus we can rearrange the order of the addends to collect the like terms.
- For Try It! 19, students may get $4 + y$, which is also correct. By convention, we usually write the constant term last so the process can be shown by grouping the variable terms first, and then the constant terms.

$$8 + 2y - 4 - y = 2y - y + 8 - 4$$
$$= y + 8 - 4$$
$$= y + 4$$

Subtraction is not commutative. After students learn how to add negative numbers in Grade 7, they will see that we can move negative terms based on the commutative property of addition, because subtracting a positive number is the same as adding a negative number.

$$8 + 2y - 4 - y = 8 + 2y + (-4) + (-y)$$
$$= 2y + (-y) + 8 + (-4)$$
$$= (2y - y) + (8 - 4)$$
$$= y + 4$$

Example 20:
- Example 20 (a):
  - Students may think that you cannot simplify these because there are variables and numbers. Point out that the multiplication sign means that it is one term. $3n + 2$ would be two terms, but $3n \times 2$ is one term. The multiplication sign is generally omitted between a coefficient and a variable in a term.
  - We can multiply in any order based on the commutative property of multiplication. Thus, $3n \times 2 = 3 \times n \times 2 = 3 \times 2 \times n = 6 \times n = 6n$.
- Example 20 (b) and Try It! 20 (b):
  - For an expression to be considered simplified, the fractions must be simplified.

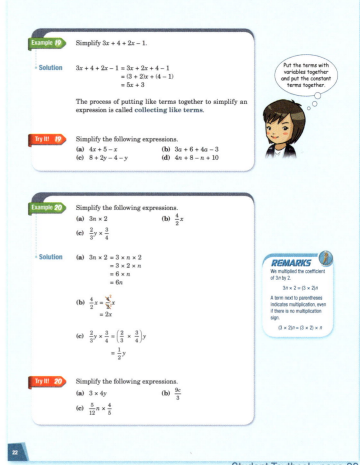

Student Textbook page 22

- Example 20 (c):
  - If students are having difficulty grasping why we can do this, it can be explained as:

$$\tfrac{2}{3}y \times \tfrac{3}{4} = \tfrac{2}{3} \times y \times \tfrac{3}{4} = \tfrac{2}{3} \times \tfrac{3}{4} \times y = \tfrac{1}{2}y$$

A similar idea can be used to explain Try It! 20 (c).

Try It! 19 Answers

(a) $4x + 5 - x = 4x - x + 5 = 3x + 5$

(b) $3a + 6 + 4a - 3 = 3a + 4a + 6 - 3 = 7a + 3$

(c) $8 + 2y - 4 - y = 2y - y + 8 - 4 = y + 4$

(d) $4n + 8 - n + 10 = 4n - n + 8 + 10 = 3n + 18$

Try It! 20 Answers

(a) $3 \times 4y = 3 \times 4 \times y = 12 \times y = 12y$

(b) $\frac{9c}{3} = \frac{9}{3}c = 3c$

(c) $\frac{5}{12}n \times \frac{4}{5} = \left(\frac{5}{12} \times \frac{4}{5}\right)n = \frac{1}{3}n$

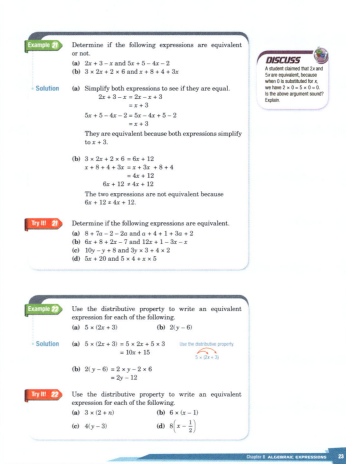

Student Textbook page 23

4. Extension

Have students study Examples 21 – 23 and do Try It! 21 – 23.

Example 21:
- We can determine whether two expressions are equivalent by simplifying both expressions. If the resulting expressions are the same, the expressions are equivalent.

DISCUSS
- This is only true for $x = 0$, which is a special case. To be equivalent expressions, they have to remain true for any value of x. If $x = 3$, for example, then $2 \times 3 \neq 5 \times 3$.

Example 22:
- Students learned the distributive property of multiplication in Chapter 1 (Lesson 1.3). Here, they are applying it to expressions that have terms with variables.
- When the expression involves multiplication with parentheses, we can use the distributive property to simplify the expression. The resulting expression is equivalent to the original expression.
- Remind students that when a number is followed immediately by a parenthesis, such as $2(y - 6)$, there is a multiplication sign that is omitted between 2 and $(y - 6)$.

Try It! 21 Answers

(a) No, $5a + 6 \neq 4a + 7$ (b) Yes, $8x + 1 = 8x + 1$

(c) Yes, $9y + 8 = 9y + 8$ (d) Yes, $5x + 20 = 20 + 5x$

Try It! 22 Answers

(a) $3 \times (2 + n) = 3 \times 2 + 3 \times n = 3n + 6$

(b) $6 \times (x - 1) = 6 \times x - 6 \times 1 = 6x - 6$

(c) $4(y - 3) = 4 \times y - 4 \times 3 = 4y - 12$

(d) $8\left(x - \frac{1}{2}\right) = 8 \times x - 8 \times \frac{1}{2} = 8x - 4$

Lesson 6

Example 23:
- Students learned about common factors and the distributive property in Chapter 1.
  Here, they are finding a common factor in each term of the expression to rewrite the expression using parentheses and the distributive property.
- Help students break down the terms based on a common factor. See the blue diagrams with arrows in solutions (a) and (b).

5. Conclusion

Summarize the main points of the lesson.
- We can simplify expressions involving addition and subtraction by combining like terms. When we do this, the simplified expression and the original expression are equivalent.
- When an expression involves multiplication with parentheses, we can use the distributive property to simplify the expression and write an equivalent expression.
- For two expressions to be equivalent, they need to have the same values for all possible numbers.

Try It! 23   Answers

(a) $6n$ and 8 have a common factor of 2.
$6n + 8 = 2 \times 3n + 2 \times 4 = 2(3n + 4)$

(b) 9 and $3x$ have a common factor of 3.
$9 - 3x = 3 \times 3 - 3 \times x = 3(3 - x)$

(c) $8y$ and 4 have a common factor of 4.
$8y + 4 = 4 \times 2y + 4 \times 1 = 4(2y + 1)$

(d) $14y$ and 21 have a common factor of 7.
$14y - 21 = 7 \times 2y - 7 \times 3 = 7(2y - 3)$

★ **Workbook: Page 14**

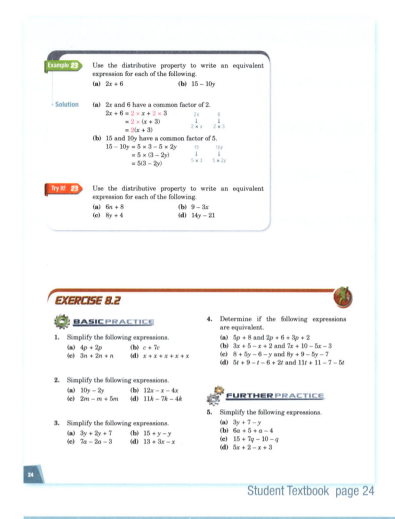
Student Textbook page 24

Objective:
- Consolidate and extend the material covered thus far.

Have students work together with a partner or in groups. Students should try to solve the problems by themselves first, then compare solutions with their partner or group. If they are confused, they can discuss together.

Observe students carefully as they work on the problems. Give help as needed individually or in small groups.

1. (a) $6p$ (b) $8c$
   (c) $6n$ (d) $5x$

28

2. (a) $8y$ (b) $7x$
   (c) $6m$ (d) 0

3. (a) $5y + 7$ (b) 15
   (c) $5a - 3$ (d) $13 + 2x$

4. (a) Equivalent (b) Equivalent
   (c) Not equivalent (d) Not equivalent

 FURTHER PRACTICE

5. (a) $2y + 7$ (b) $7a + 1$
   (c) $5 + 6q$ (d) $4x + 5$

6. (a) Equivalent (b) Not equivalent
   (c) Equivalent (d) Not equivalent

7. (a) $10 + 5r$ (b) $4x - 8$
   (c) $0.5x - 2$ (d) $\frac{2}{3}y + 4$

8. (a) $5(x + 1)$ (b) $3(1 - 2y)$
   (c) $3(2x + 5)$ (d) $5(2y - 1)$

 MATH@WORK

9. $2(l + w) = 2(12 + 4) = 32$; 32 in

10. (a) $2(4p + 2p) = 2(6p) = 12p$; $12p$ cm
    (b) $12 \times 8 = 96$; 96 cm

11. (a) $5x + 4x + 12 = 9x + 12$, $(9x + 12)$ m
    (b) $9 \times 4 + 12 = 48$; 48 m
    (c) Students should draw triangles with the following dimensions.
    When $x = 3$; 12 m, 15 m, 12 m
    When $x = \frac{12}{5}$; 12 m, 12 m, $9\frac{3}{5}$ m
    When $x = 1$; 12 m, 5 m, 4 m

12. Now, Stephanie is y years old and Kalama is $2y$ years old. In 2 years, Stephanie will be $(y + 2)$ years and Kalama will be $(2y + 2)$ years. In 2 years, Kalama will not still be twice as old as Stephanie because $2(y + 2) \neq 2y + 2$. Note: The answer in the back of the student book incorrectly names Kalama as Jacqueline.

BRAIN WORKS

13. Possible answers: $5x - x + 5 - 13$ or $4(x - 2)$

14. $2(l + w) = 2 \times l + 2 \times w = 2l + 2w$

15. Let b denote the length of the pole which was painted blue, w for white, r for red, and L for the length of the flagpole. $r + w + b = L$ is known. The sum of lengths of the blue and the white parts is $(b + w)$, and the sum of lengths of the red and the white parts is $(r + w)$. So the length of the pole which is painted white is
$w = (b + w) + (r + w) - L.$

Lesson 7

Objectives:

- Summarize and reflect on important ideas learned in this chapter.
- Apply and extend understanding by investigating a non-routine problem.

Student Textbook page 26

Note: This lesson could be done in class or assigned for students to do independently at home or at school.

1. **In a Nutshell**

Use this page to summarize the important ideas learned in this chapter.

Give examples where needed.

2. Write In Your Journal

Have students reflect on and do the writing activity. If students are confused, ask them to draw a bar model to understand why it is $S = 6T$, not $6S = T$.

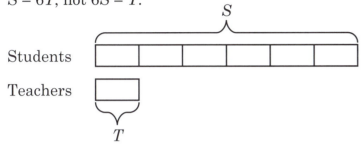

3. Extend Your Learning Curve

Have students work on this problem with partners or in groups and share the rectangles they found with the class.

Students can explore different rectangles by trial and error initially. It may be helpful to fix one dimension. For example, try 3 by 4, then 3 by 5 ... they are getting closer, so try 3 by 6.

Squares are also rectangles. To think about it systematically, have students think about the areas and perimeters of squares by making a table.

Side length of square	Area	Perimeter
1	1	4
2	4	8
3	9	12
4	16	16
5	25	20
6	36	24

Notes:
- The area and perimeter of a rectangle will be numerically equal when the product of the sides is twice as much as the sum of the sides, $ab = 2(a + b)$.

- This is an informal investigation. Students have not learned how to manipulate equations yet but this can be understood algebraically as:

$$ab = 2(a + b)$$
$$ab = 2a + 2b$$
$$ab - 2a = 2b$$
$$a(b - 2) = 2b$$
$$a = \frac{2b}{b - 2}$$

Substitute any number for b in the last equation and you will get the dimension needed for a. If $b = 10$,

$$a = \frac{2 \times 10}{10 - 2} = \frac{20}{8} = 2.5$$

Thus, a rectangle with dimensions of 10 and 2.5 will also work.

continues on next page ...

Write in Your Journal

Possible answers

- $S = 6T$
- Her answer is incorrect. $6S = T$ means that there are 6 times as many teachers as students.

Extend Your Learning Curve

- The area and perimeter of a 6 by 3 rectangle are numerically equal, but area measures the space inside the rectangle and perimeter measures the length around the outside of the rectangle.

Possible answers:

- 4 by 4
- 10 by 2.5

Chapter 9: Equations and Inequalities

Lesson	Objectives	Class Periods	Textbook & Workbook	Teacher's Guide Page	Additional Materials Needed
1	• Determine whether a given number makes an equation true.	1	TB: 28–31 WB: 19–21	38	
2	• Balance simple one-step equations and understand their properties. • Solve simple algebraic equations.	1	TB: 32–39	42	Balance scale, connecting cubes (2 colors), small stickers for labels
3	• Solve word problems using algebraic equations.	1	TB: 39–42 WB: 22–33	50	
4	• Consolidate and extend the material covered thus far.	1	TB: 42–43	54	
5	• Determine whether a given number makes an inequality true.	1	TB: 44–45 WB: 34	58	Balance scale, connecting cubes (2 colors), small stickers for labels
6	• Graph inequalities using a number line.	1	TB: 46–49	60	Graph paper, rulers
7	• Consolidate and extend the material covered thus far.	1	TB: 49–50 WB: 35–39	63	
8	• Summarize and reflect on important ideas learned in this chapter. • Use a spreadsheet to evaluate expressions and solve equations.	1	TB: 51–52	65	
9	• Apply the model method to solve complex word problems involving algebraic equations.	1	TB: 53–56	67	

Chapter 9 **EQUATIONS AND INEQUALITIES** 33

2017 Singapore Math Inc. Dimensions Math® Teacher's Guide 6B

Chapter 9: Equations and Inequalities

In elementary school, students learned how to solve equations where a ☐ represents the unknown quantity based on part-whole relationships.

In additive situations:

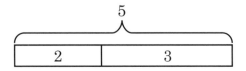

To find an unknown total, we add the parts.

2 + 3 = ☐ → ☐ = 2 + 3

☐ − 3 = 2 → ☐ = 2 + 3

To find an unknown part, we subtract the known part from the total.

2 + ☐ = 5 → ☐ = 5 − 2

5 − ☐ = 2 → ☐ = 5 − 2

In multiplicative situations:

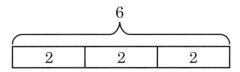

To find an unknown total, we multiply the factors.

3 × 2 = ☐ → ☐ = 3 × 2

☐ ÷ 3 = 2 → ☐ = 2 × 3

☐ ÷ 2 = 3 → ☐ = 3 × 2

To find the value of an unknown factor, we divide the total by the known factor.

2 × ☐ = 6 → ☐ = 6 ÷ 2

☐ × 3 = 6 → ☐ = 6 ÷ 3

6 ÷ ☐ = 3 → ☐ = 6 ÷ 3

6 ÷ ☐ = 2 → ☐ = 6 ÷ 2

In this chapter, these ideas are extended to solving algebraic equations. Instead of ☐, a variable represents the unknown quantity in the equations.

Chapter 9: Equations and Inequalities

Initially, students can solve algebraic equations by inspection. In the equation $x + 8 = 12$, students can know that $x = 4$ because they know that $4 + 8 = 12$. Students can then substitute 4 for x to see whether the value is the solution to the equation.

$x + 8 = 12$

$4 + 8 = 12$

$12 = 12$

An equation is a statement of equality between two expressions, thus the expressions on both sides of the equal sign are equivalent. When we perform the same operation to both sides of an equation, the expressions on each side of the equal sign remain equivalent.

$x + 8 = 12$ $(x + 8) - 4 = 12 - 4$ $2(x + 8) = 2 \times 12$ $(x + 8) \div 2 = 12 \div 2$

We can use this idea to solve equations. For example,

$x + 8 = 12$

$x + 8 - 8 = 12 - 8$

$x = 4$

The **solution** to an equation is the number that makes the equation true. It is important to encourage students to check their answers by substituting the solution for the variable in the original equation.

$4 + 8 = 12$

$12 = 12$

Bar models are useful to help students make the connection between arithmetic and algebra. With algebra, they use a letter instead of a question mark to indicate the unknown.

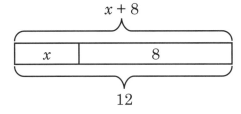

Chapter 9: Equations and Inequalities

An **inequality** is a statement that indicates that one quantity is less than or greater than another. Students learned about inequalities when they compared numbers using > and < in elementary school. In Chapter 8, they learned the inequality sign ≠ (not equal to). In this chapter, students will learn the inequality signs, ≥ (greater than or equal to), and ≤ (less than or equal to).

Unlike an equation, an inequality can have an infinite number of values that satisfy the inequality. However, as with equations, substitution can be used to verify whether a specific value satisfies the inequality, that is, whether it is "a solution" to the inequality.

In the inequality $x > 2$, the solution includes every number greater than 2. 3 is a solution for $x > 2$ because 3 is greater than 2. −3 is not a solution for $x > 2$ because −3 is not greater than 2.

We can graph the solution to an inequality using a number line.

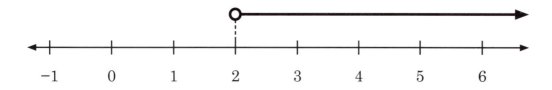

The open circle signifies that 2 is not included in the solution to $x > 2$.

A closed circle is used to show that the value being compared is included in the solution, such as $x \geq 2$.

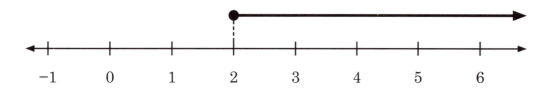

In **Dimensions Math® Textbook 6B**, students will only be determining whether a value is a solution to an inequality and graphing the solution set. They will not be solving inequalities involving algebraic expressions, such as $3x + 4 < 32$.

Notes

Lesson 1

Objective:

- Determine whether a given number makes an equation true.

1. Introduction

Read and discuss the Chapter Opener and state the learning goals of the chapter.

Write ☐ + ☐ = ☐ and ask students what numbers can go in the ☐s.

Write ☐ + ☐ = 12 and ask students what numbers can go in the ☐s.

Write ☐ + 8 = 12 and ask students what number can go in the ☐.

Write $x + 8 = 12$ and ask students what the value of x is.

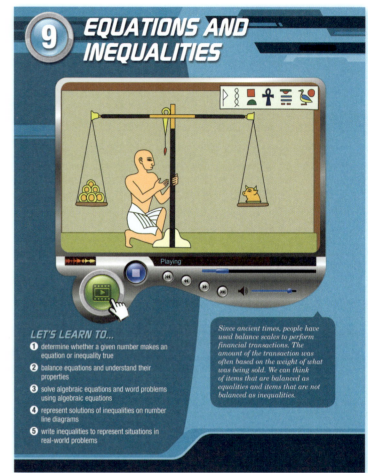

Student Textbook page 28

Notes:

- Point out that on each side of the equation there is an expression. E.g., in the equation $x + 8 = 12$, $x + 8$ is an expression and 12 is also an expression. In an equation, the two expressions are equivalent (see REMARKS).
- The solution is the number that, when substituted for the variable, makes the equation true. The only solution that will make the equation $x + 8 = 12$ true is $x = 4$.
- If we add or subtract the same number to both sides of the equation, the expressions remain equivalent. E.g., in $x + 8 = 12$, if we substitute 4 for x and subtract 8 from each side, the expressions will have the same value; $4 + 8 - 8 = 12 - 8$.
- Point out how the equal signs are aligned in the solution process.

2. Development

Have students try Example 1 on their own (without looking at the solution) and share their solutions and methods. Discuss the solutions on page 30.

Have students do Try It! 1 on their own and discuss the solutions.

Student Textbook page 29

Try It! 1 Answers

(a) $5 + 18 = 23$, Yes (b) $12 - 5 \neq 6$, No

(c) $26 = 31 - 5$, Yes (d) $5 + \frac{2}{3} = \frac{17}{3}$, Yes

3. **Application**

Have students study Examples 2 – 3 and do Try It! 2 – 3.

Notes:
- In Example 2 (b) and Try It! 2 (c) and 2 (d), the coefficients are fractions or decimals. In Example 3 and Try It! 3, the value of the variable is a fraction. Look for mechanical errors and provide help as needed.

Try It! 2   Answers

(a) $6 \times 8 \neq 40$, No

(b) $96 = 12 \times 8$, Yes

(c) $\frac{1}{2} \times 8 = 4$, Yes

(d) $1.5 \times 8 = 12$, Yes

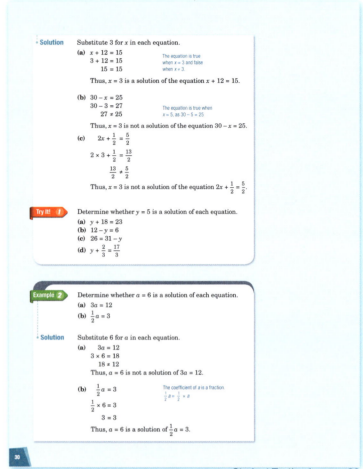

Student Textbook page 30

Notes:

- For problems involving fractions, make sure that students simplify the fractions on each side of the equation to determine whether they are equivalent. In Try It! 3 (d), for example, students may do $\frac{3}{5} \times \frac{2}{3} = \frac{6}{15}$ and not realize that $\frac{6}{15} = \frac{2}{5}$.

- For Try It! 3 (c), students should recognize that $\frac{3}{2}$ and $\frac{2}{3}$ are reciprocals.

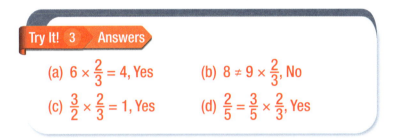

Try It! 3 Answers

(a) $6 \times \frac{2}{3} = 4$, Yes   (b) $8 \neq 9 \times \frac{2}{3}$, No

(c) $\frac{3}{2} \times \frac{2}{3} = 1$, Yes   (d) $\frac{2}{5} = \frac{3}{5} \times \frac{2}{3}$, Yes

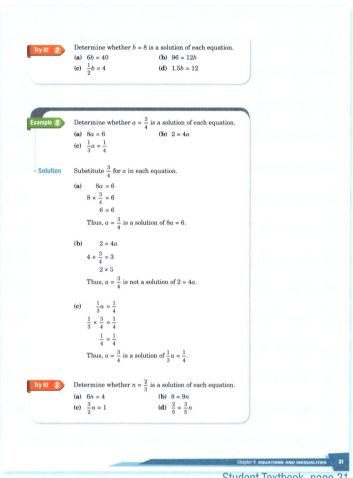

Student Textbook page 31

4. Conclusion

Summarize important points of the lesson.
- The expressions on each side of an equation must be equivalent.
- We can substitute a value for a variable to determine if it is a solution to the equation.

★ **Workbook: Page 19**

Lesson 2

Objectives:

- Balance simple one-step equations and understand their properties.
- Solve simple algebraic equations.

1. Introduction

Read and discuss page 32.

Note:

Here, students are just noticing that if we add, subtract, multiply, or divide both sides of the equation by the same number, the equation remains true. They will prove this in the Class Activity on the following page.

B **Balancing Equations**

We have learned that in an equation, the expressions on both sides of the equal sign must be equivalent.

For example,

$$3 + 4 = 14 \div 2 \qquad \text{because } 7 = 7$$
$$4 \times 6 - 3 = 3 \times 7 \qquad \text{because } 21 = 21$$
$$3x + 3 + 5 = x + 8 + 2x \quad \text{because } 3x + 8 = 3x + 8$$

If the expressions on both sides are not equivalent, it is not an equation.

For example,

$$3 \times 4 \neq 5 + 8 \qquad \text{because } 12 \neq 13$$
$$8 \times 2 + 3 \neq 3^2 + 1 \qquad \text{because } 19 \neq 10$$
$$2y + 3 - y \neq y + 10 - 6 \quad \text{because } y + 3 \neq y + 4$$

If we add the same number to, or subtract the same number from, both sides of an equation, the expressions on both sides remain equivalent.

Similarly, if we multiply or divide both sides of an equation by the same number, the expressions on both sides remain equivalent. For example,

If $\qquad 4 + 2 = 2 \times 3,$

then
$$(4 + 2) + 4 = (2 \times 3) + 4$$
$$(4 + 2) - 1 = (2 \times 3) - 1$$
$$(4 + 2) \times 2 = (2 \times 3) \times 2$$
$$(4 + 2) \div 3 = (2 \times 3) \div 3$$

If we add, subtract, multiply, or divide each side by different numbers, the expressions on each side will not be equivalent. For example,

If $\qquad 4 + 2 = 2 \times 3,$

then
$$(4 + 2) + 4 \neq (2 \times 3) + 3$$
$$(4 + 2) - 2 \neq (2 \times 3) - 1$$
$$(4 + 2) \times 2 \neq (2 \times 3) \times 3$$
$$(4 + 2) \div 3 \neq (2 \times 3) \div 5$$

32

Student Textbook page 32

2. Development

Give students the materials for Class Activity 1 and have them do the activity with partners or in groups.

Answers for Class Activity 1

3. $x + 2 = 3 + \boxed{2}$

4. (a) It drops to the right, because $x = 3$ and $3 < 5$.

   (b) It becomes equal because $x = 3$ and $3 = 3$.

   (c) $x + 2 - 2 = 3 + 2 - \boxed{2}$

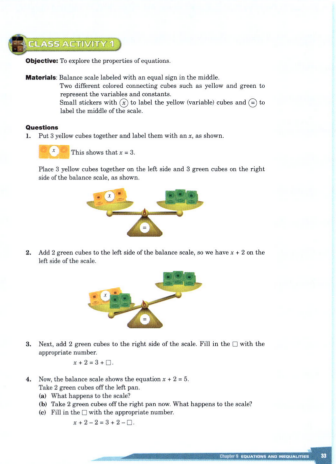

Student Textbook page 33

Answers for Class Activity 1 (continued)

5. It remains balanced because (the weight of) each side is equal.

6. It is not balanced because (the weight of) each side is not equal.

8. (a) Since $x = 3$, we need to add 6 more cubes.

    (b) $2x \times 2 = 6 \times \boxed{2}$

9. (a) Take off 6 cubes.

    (b) $\dfrac{4x}{2} = \dfrac{12}{\boxed{2}}$

Ask students to share their conclusions and then summarize the results of the activity from the top of page 35.

3. Application

Have students study Examples 4 – 9 and do Try It! 4 – 9.

Notes:
- There are several methods presented for each problem. Students can use the method of their choice, but they should understand each method and discuss the merits of each.
- When solving equations, make sure students align the equal signs and check their answers.
- Students may be able to solve many equations simply by inspection. E.g., in $p + 7 = 23$, students may know $p = 16$ because $16 + 7 = 23$. Encourage students to show their thinking algebraically.
- Method 1 uses the idea of balancing the equation and includes all of the steps needed to solve the equation.
- Method 2 is related to students' prior knowledge of the part-whole relationship of addition and subtraction. In this case, we know the total (36) and one part (17). To find the missing part, we subtract the known part from the total (see REMARKS). Based on this idea, we can just rewrite $x + 17 = 36$ as $x = 36 - 17$. Students who are having difficulty with this method should be encouraged to draw a model.

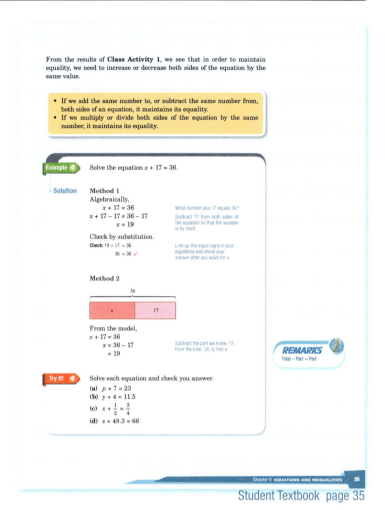

Student Textbook page 35

Try It! 4 Answers

(a) $p + 7 = 23$
$p + 7 - 7 = 23 - 7$
$p = 16$

(b) $y + 4 = 11.5$
$y + 4 - 4 = 11.5 - 4$
$y = 7.5$

(c) $x + \frac{1}{2} = \frac{3}{4}$
$x + \frac{1}{2} - \frac{1}{2} = \frac{3}{4} - \frac{1}{2}$
$x = \frac{1}{4}$

(d) $s + 48.3 = 66$
$s + 48.3 - 48.3 = 66 - 48.3$
$s = 17.7$

Note:
- In Example 5, Method 2, we know the two parts but we do not know the total. To find the total, we add the two parts (see REMARKS). Based on this idea, we can rewrite $x - 8 = 14$ as $x = 14 + 8$.

Try It! 5 Answers

(a) $n - 5 = 12$
$n - 5 + 5 = 12 + 5$
$n = 17$

(b) $y - 87 = 56$
$y - 87 + 87 = 56 + 87$
$y = 143$

(c) $x - \frac{1}{3} = \frac{1}{6}$
$x - \frac{1}{3} + \frac{1}{3} = \frac{1}{6} + \frac{1}{3}$
$x = \frac{1}{2}$

(d) $r - 0.75 = 1.25$
$r - 0.75 + 0.75 = 1.25 + 0.75$
$r = 2$

Notes:
- In Example 6, the variable is on the right side of the equation. Students who are having difficulty could rewrite the equation as $x + 19 = 48$.
- In algebra, the solution is usually given with the answer on the right side of the equation (see bottom REMARKS on page 36).

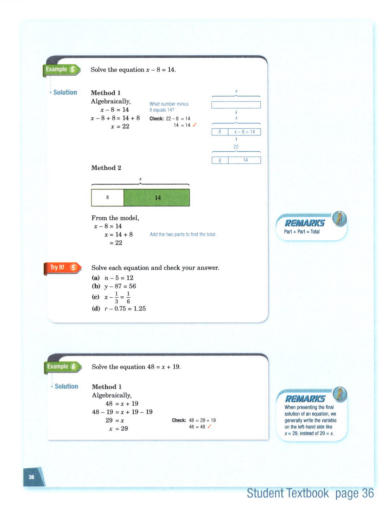

Student Textbook page 36

- Method 2: The first step shown is to rewrite the equation with the variable on the left side (see REMARKS on page 37). It could also be solved without rewriting the equation as:
$48 = x + 19$
$48 - 19 = x$
$x = 29$
- When solving equations, we can leave an answer as an improper fraction or change it to a mixed number, whichever is easiest. Thus, Try It! 6 (b) could also be expressed as $x = \frac{11}{3}$ or $x = 3\frac{2}{3}$, and Try It! 6 (d) could be expressed as $x = 2\frac{3}{4}$ or $x = \frac{11}{4}$. When solving word problems with equations, however, the answer should be expressed the way that makes the most sense.

Try It! 6 Answers

(a) $43 = m + 18$
$43 - 18 = m + 18 - 18$
$m = 25$

(b) $5 = x + \frac{4}{3}$
$5 - \frac{4}{3} = x + \frac{4}{3} - \frac{4}{3}$
$x = 3\frac{2}{3}$

(c) $35 = y - 17$
$35 + 17 = y - 17 + 17$
$y = 52$

(d) $\frac{3}{4} = x - 2$
$\frac{3}{4} + 2 = x - 2 + 2$
$x = 2\frac{3}{4}$

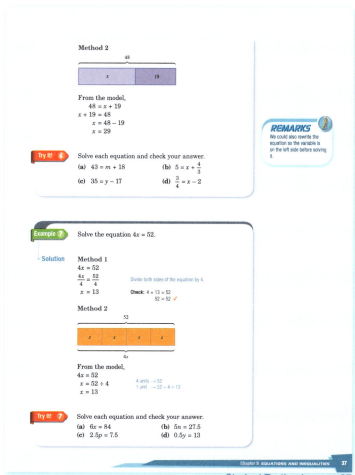

Student Textbook page 37

Notes:
- Since the left side of the equation in Example 7 is $x \times 4$, to isolate the variable we divide each side by 4. In Method 1, this is done using a fraction. Remind students that $\frac{4x}{4} = \frac{4}{4} \times x = 1 \times x = x$.
- In Method 2, the bar model and unitary methods are shown, as these are familiar to students. 4 units = 52, one unit = 52 ÷ 4. Thus, when solving the equation, we can rewrite $4x = 52$ as $x = 52 \div 4$.
- For Try It! 7, encourage students who are struggling to draw a bar model.

Try It! 7 Answers

(a) $6x = 84$
$\frac{6x}{6} = \frac{84}{6}$
$x = 14$

(b) $5n = 27.5$
$\frac{5n}{5} = \frac{27.5}{5}$
$n = 5.5$

(c) $2.5p = 7.5$
$\frac{2.5p}{2.5} = \frac{7.5}{2.5}$
$p = 3$

(d) $0.5y = 13$
$\frac{0.5y}{0.5} = \frac{13}{0.5}$
$y = 26$

4. Extension

Have students study Examples 8 – 9 and do Try It! 8 – 9.

Notes:

- These examples deal with fractional coefficients. To isolate the variable, we multiply the fraction by its reciprocal. If needed, review the idea of multiplying a fraction by its reciprocal to get 1.

- In Example 8, $\frac{x}{3} = \frac{1}{3}x$, so to isolate the variable, multiply $\frac{x}{3}$ by the reciprocal of $\frac{1}{3}$, which is 3.

- Multiplication and division are inverse operations. If we divide a number by 3, and then multiply it by 3, we will get the same number.
$$x \div 3 = 5$$
$$x \div 3 \times 3 = 5 \times 3$$
$$x = 15$$

- In Method 2, the bar model shows that the total (whole) is x, and there are 3 parts. Thus one unit (part) is $x \div 3 = \frac{x}{3}$. Since 1 unit = 5, 3 units = 3×5.

- Encourage students who are having difficulty to draw a bar model.

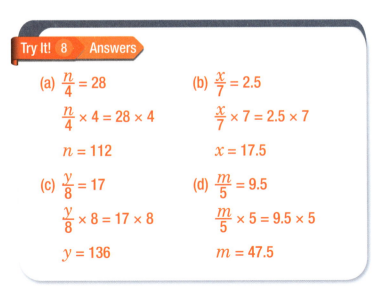

Try It! 8   Answers

(a) $\frac{n}{4} = 28$ (b) $\frac{x}{7} = 2.5$

$\frac{n}{4} \times 4 = 28 \times 4$ $\frac{x}{7} \times 7 = 2.5 \times 7$

$n = 112$ $x = 17.5$

(c) $\frac{y}{8} = 17$ (d) $\frac{m}{5} = 9.5$

$\frac{y}{8} \times 8 = 17 \times 8$ $\frac{m}{5} \times 5 = 9.5 \times 5$

$y = 136$ $m = 47.5$

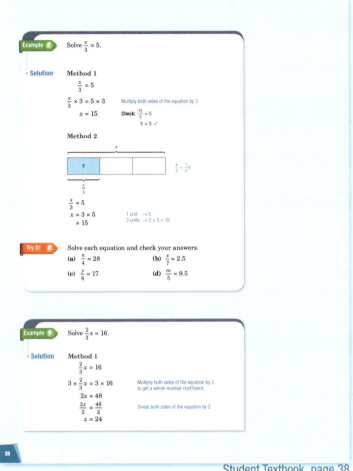

Student Textbook page 38

Notes:

- In Example 9, Method 1 involves simplifying the calculation by multiplying by the denominator of the coefficient to get a whole number coefficient. This method is important because in the future, students will use this idea to eliminate the fractions in more complex equations. For example, in $\frac{2}{3}x + \frac{1}{2} = \frac{1}{3}$, we can multiply both sides of the equation by 6 to eliminate the fractions.

- $\frac{2}{3}x = \frac{2x}{3} = 2x \div 3$, thus we could solve it as:
$$2x \div 3 = 16$$
$$2x = 16 \times 3$$
$$2x = 48$$
$$x = 24$$

48   ©2017 Singapore Math Inc. Dimensions Math® Teacher's Guide

- Method 2 uses a bar model to illustrate the unitary method. When we divide the total (x) by 3, each unit is $\frac{1}{3}x$, so 2 units = $\frac{2}{3}x$. Since $\frac{2}{3}x = \frac{1}{3}x \times 2$, we can divide the right side of the equation by 2 and then multiply it by 3.

This can also be shown as:

$\frac{2}{3}x = 16$

$\frac{1}{3}x = 16 \div 2$

$\frac{1}{3}x = 8$

$x = 8 \times 3$

$x = 24$

Encourage students who are having difficulty with this method to draw a bar model.

- Method 3 is perhaps the simplest method, because we can isolate the variable in one step by multiplying each side by the reciprocal of the coefficient.

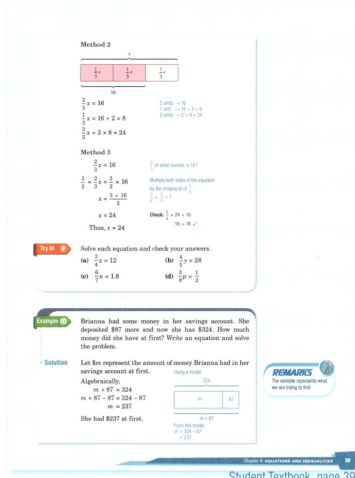

Student Textbook page 39

5. Conclusion

Summarize the important points of the lesson.
- If we add, subtract, multiply, or divide each side of the equation by the same number it will maintain its equality.
- We can use this idea to solve algebraic equations by isolating the variable on one side of the equation.

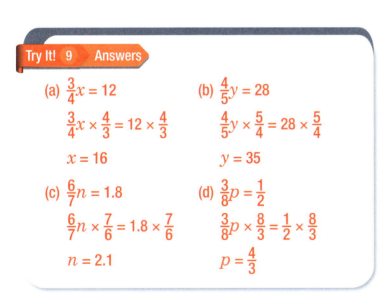

Try It! 9 Answers

(a) $\frac{3}{4}x = 12$

$\frac{3}{4}x \times \frac{4}{3} = 12 \times \frac{4}{3}$

$x = 16$

(b) $\frac{4}{5}y = 28$

$\frac{4}{5}y \times \frac{5}{4} = 28 \times \frac{5}{4}$

$y = 35$

(c) $\frac{6}{7}n = 1.8$

$\frac{6}{7}n \times \frac{7}{6} = 1.8 \times \frac{7}{6}$

$n = 2.1$

(d) $\frac{3}{8}p = \frac{1}{2}$

$\frac{3}{8}p \times \frac{8}{3} = \frac{1}{2} \times \frac{8}{3}$

$p = \frac{4}{3}$

Lesson 3

Objective:

- Solve word problems using algebraic equations.

1. Introduction

Pose the following problem: Brianna had some money in her savings account. She deposited $20 more and now she has $50. How much money did she have at first?

Note: This is the same problem as Example 10, but with simplified numbers.

Tell students to let m = the money she had at first and ask them to draw a bar model to represent the situation.

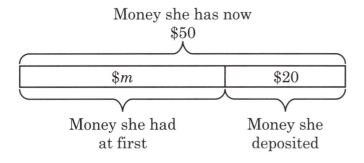

Ask students to write an equation to represent the problem ($m + 20 = 50$).

Pose the problem with the same numbers as Example 10 ($87 and $324), and ask them to represent the problem with an equation, then use the equation to solve the problem.

Have students share their methods for solving the problem, then discuss the method shown for Example 10 on page 39. Discuss REMARKS.

Have students check their answers by substituting $237 for m.

Note: Students who have learned bar models before should be able to see that to find the value of m, we simply subtract 20 from 50.

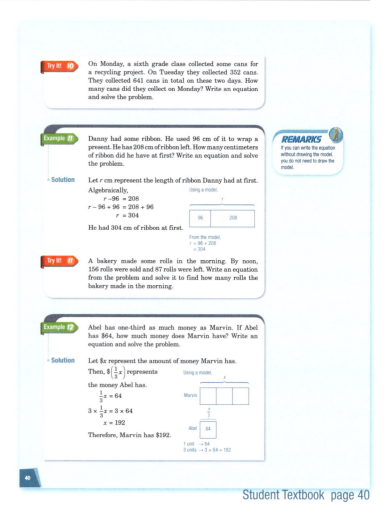

Student Textbook page 40

The emphasis here should not be on finding the answer, but on using an equation to model the problem situation. The bar model is helpful to see the equation visually.

Have students solve Try It! 10 by drawing a model and writing an equation. Ask them to share their methods and discuss.

Try It! 10 Answer

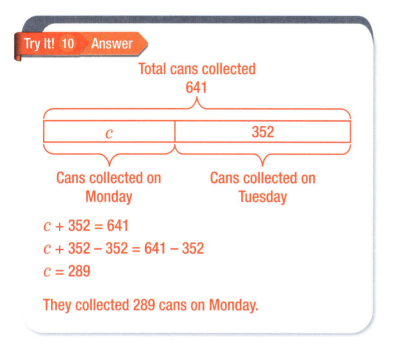

$c + 352 = 641$
$c + 352 - 352 = 641 - 352$
$c = 289$

They collected 289 cans on Monday.

Notes:
- Based on the model, students may also write $641 - 352 = c$ or $c = 641 - 352$. These equations are acceptable, but it is usually best for the equation to mirror the order of the problems. I.e., "She started with some cans (c). She collected 352 more cans ($+ 352$). Then she had a total of 641 cans ($= 641$)."
- Students can assign any variable for the unknown. Initially, it is helpful to assign the first letter of what we are trying to find. For example, "We are finding cans, so we can use c to represent the cans collected on Monday." Later, students can use x to represent any unknown quantity (see Examples 13 – 14).
- Encourage students to check the answer by substituting 289 for c in the original equation.

2. Application

Have students study Examples 11 – 13 and do Try It! 11 – 13.

After students have solved the problems, have them share their equations and models. Relate the quantities in the models to the equations.

Notes:
- Encourage students who are having difficulty to draw a model first and then write an equation.
- Students who do not need to draw a model do not have to (see REMARKS).
- In Example 11, the total is unknown but we know both parts. Remind students that part + part = total. The model shows this clearly.

Try It! 11 Answer

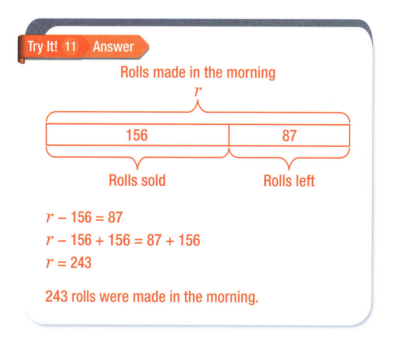

$r - 156 = 87$
$r - 156 + 156 = 87 + 156$
$r = 243$

243 rolls were made in the morning.

Notes:

- In Example 12, we are comparing Marvin's money to Abel's money. The value being compared to is the base, so Marvin's money is $1x = x$. Abel's money is $\frac{1}{3}$ of that or $\frac{1}{3}x$.

- For Try It! 12, students may write the equation $s = 267 \times 4$. This is fine, but help students see that we can express the number of local stamps in terms of the number of foreign stamps. The local stamps become $\frac{1}{4}s$.

Try It! 12 Answer

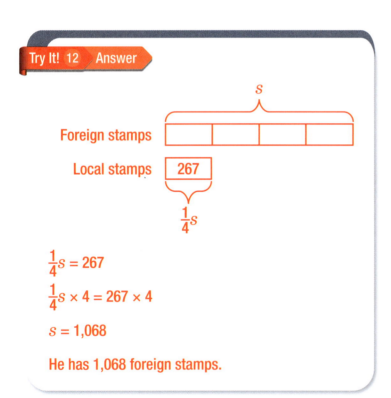

$\frac{1}{4}s = 267$

$\frac{1}{4}s \times 4 = 267 \times 4$

$s = 1,068$

He has 1,068 foreign stamps.

Try It! 13 Answer

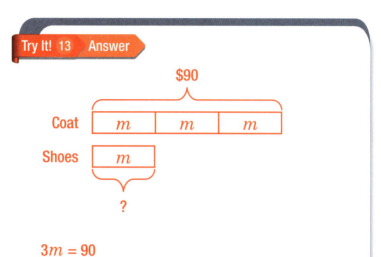

$3m = 90$

$\frac{3m}{3} = \frac{90}{3}$

$m = 30$

He spent $30 on the shoes.

3. **Extension**

Have students study Example 14 and do Try It! 14. Then, have them share their solutions.

Notes:

- We are comparing the price of the sweater to the price of the coat, so the coat is the base ($1x$). Since the sweater costs $\frac{3}{5}$ as much, we can represent the price of the sweater as $\frac{3}{5}x$.

- From the bar model, we can see that the coat is 5 units. Since all 5 units are x, one unit is $\frac{x}{5}$ or $\frac{1}{5}x$.

- Students learned both methods in Example 9. Method 2 involves fewer steps.

★ **Workbook: Page 22**

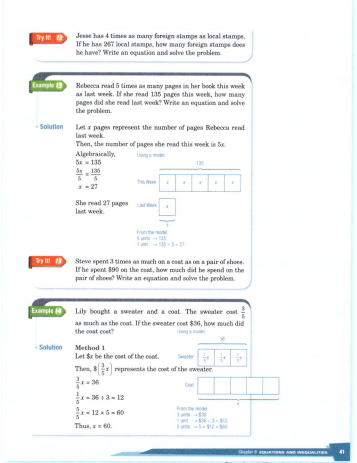

Student Textbook page 41

Lesson 4

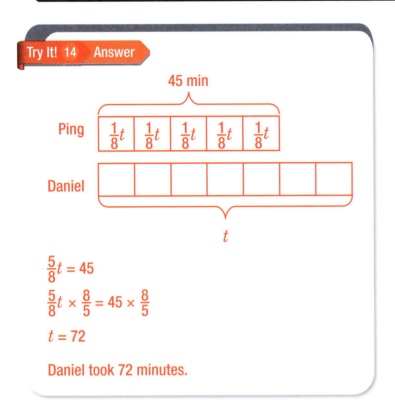

Try It! 14 Answer

Ping, Daniel bar model with 45 min

$\frac{5}{8}t = 45$

$\frac{5}{8}t \times \frac{8}{5} = 45 \times \frac{8}{5}$

$t = 72$

Daniel took 72 minutes.

4. **Conclusion**

Summarize the main points of the lesson.
- We can represent and solve problem situations with algebraic equations.
- To think about the equation needed to solve a problem, it is often helpful to draw a bar model. When we learned bar models, we used a question mark to indicate what we were trying to find out (the unknown). With algebra, we use a variable instead of a question mark.

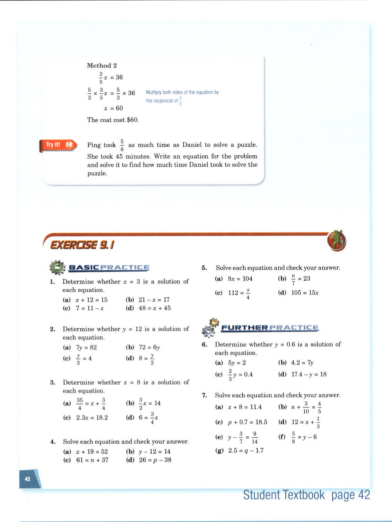

Student Textbook page 42

Objective:
- Consolidate and extend the material covered thus far.

Have students work together with a partner or in groups. Students should try to solve the problems by themselves first, then compare solutions with their partner or group. If they are confused, they can discuss together.

Observe students carefully as they work on the problems. Give help as needed individually or in small groups.

1. (a) Yes   (b) No
   (c) No   (d) Yes

2. (a) No (b) Yes
   (c) Yes (d) No

3. (a) Yes (b) No
   (c) No (d) Yes

4. (a) $x = 33$ (b) $y = 26$
   (c) $n = 24$ (d) $p = 64$

5. (a) $x = 13$ (b) $n = 161$
   (c) $x = 448$ (d) $x = 7$

FURTHER PRACTICE

6. (a) No (b) Yes
   (c) Yes (d) No

7. (a) $x = 3.4$ (b) $n = \frac{1}{2}$
   (c) $p = 17.8$ (d) $x = 11\frac{2}{3}$
   (e) $y = \frac{15}{14}$ (f) $y = 6\frac{5}{8}$
   (g) $q = 4.2$

8. (a) $n = 2.8$ (b) $x = 10$
   (c) $w = 2$ (d) $y = \frac{20}{3}$
   (e) $n = 5$ (f) $y = 2$
   (g) $x = 17$

MATH@WORK

9. $m + 150 = 325$, $m = 175$

10. $2r = 128$, $r = 64$

11. $\frac{2}{3}d = 10$, $d = 15$

12.
    Big tank | w | w | w
    Small tank | w |
    ?
    74
    $3w - w = 74$
    $2w = 74$
    $w = 37$
    2 units → 74
    1 unit → 74 ÷ 2 = 37
    37 gallons

13.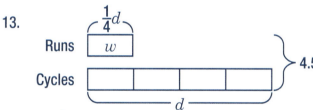
    Runs | w
    Cycles
    $\frac{1}{4}d$
    d
    4.5
    $d + \frac{1}{4}d = 4.5$
    $\frac{5}{4}d = 4.5$
    $\frac{5}{4}d \times \frac{4}{5} = 4.5 \times \frac{4}{5}$
    $d = 3.6$
    3.6 miles

    5 units → 4.5
    1 unit → 4.5 ÷ 5 = 0.9
    4 units → 0.9 × 4 = 3.6

14. Note: This is a two-step problem. Students did not see two-step equations in previous lessons, so it is important to discuss it here. Help them see that first we subtract 18 from each side of the equation, and then divide each side of the equation by 2.

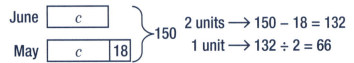

2 units $\rightarrow$ 150 − 18 = 132
1 unit $\rightarrow$ 132 ÷ 2 = 66

$c + (c + 18) = 150$
$2c + 18 = 150$
$2c = 132$
$c = 66$

66 cakes

15.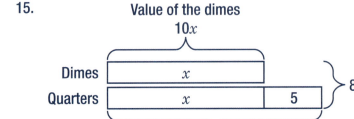

Let x = Number of dimes
$x + 5$ = Number of quarters
$10x$ = Value of the dimes (in cents)
$25(x + 5)$ = Value of quarters (in cents)
825 = Total amount of money (in cents)

$10x + 25(x + 5) = 825$
$10x + 25x + 125 = 825$
$35x + 125 = 825$
$35x = 700$
$x = 20$

There are 20 dimes.

Note: The equation could be written in dollars as $0.10x + 0.25(x + 5) = 8.25$, but it is easier to eliminate the decimals and just think about it with whole numbers using cents. It is the same as multiplying each side of the equation by 100.

$100(0.10x + 0.25(x + 5)) = 100(8.25)$
$10x + 25(x + 5) = 825$

BRAIN WORKS

16.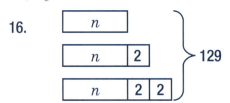

Note:
All odd numbers are 2 apart from each other. E.g., $1 + 2 = 3, 3 + 2 = 5, 5 + 2 = 7$, etc. so we can represent the numbers as $n, n + 2$ and $n + 4$.

This problem can be solved using the unitary method as:
3 units $\rightarrow$ 129 − 6 = 123
1 unit $\rightarrow$ 123 ÷ 3 = 41
So the numbers are 41, 43, and 45.
$41 \times 43 \times 45 = 79{,}335$

$n + (n + 2) + (n + 4) = 129$
$3n + 6 = 129$
$3n = 123$
$n = 41$
$n + 2 = 43$
$n + 4 = 45$
$41 \times 43 \times 45 = 79{,}335$
Be advised that the answer in the back of the textbook on page 249 is listed incorrectly as 79,355.

17. (a) $8x + 3x = 11x$
    (b) $11y - 4y = 7y$

18. Note: This is also a two-step equation and should be discussed.

    The student should have added 20 to each side of the equation before dividing each side by 2.

    $4x - 20 = 16$
    $4x = 16 + 20$
    $4x = 36$
    $x = \frac{36}{4}$
    $x = 9$

    Or, he should have divided 20 by 4 as well.

    $4x - 20 = 16$
    $(4x - 20) \div 4 = 16 \div 4$
    $x - 5 = 4$
    $x = 9$

19.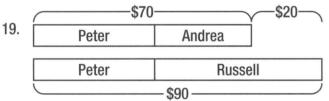

    If we compare Peter and Russell to Peter and Andrea, we see that Russell has $90 - $70 = $20 more than Andrea.

    Let a = Andrea's money
    $a + 20$ = Russell's money

    Andrea and Russell together have $80, so …

    $a + (a + 20) = 80$
    $2a + 20 = 80$
    $2a = 60$
    $a = 30$

    Andrea has $30.
    Russell has $30 + $20 = $50.
    Since Andrea and Peter have $70 together, Peter's amount is $70 − $30 = $40.
    Together they have $30 + $50 + $40 = $120, so yes, they have exactly enough money.

Check:
Peter + Andrea = $40 + $30 = $70
Andrea + Russell = $30 + $50 = $80
Peter + Russell = $40 + $50 = $90

Notes:
- This is just one of several ways to solve this problem. For example, you could compare Andrea and Peter to Andrea and Russell and see that Russell has $10 more than Peter, so …

    $r + (r - 10) = 90$
    $2r - 10 = 90$
    $2r = 100$
    $r = 50$

    Russell has $50. Peter has $50 − $10 = $40, etc.
- The problem can also be solved simply by reasoning:
    Peter + Andrea = $70
    Andrea + Russell = $80
    Peter + Russell = $90

    Each person's name appears twice, so the sum of the amounts is twice as much as the amount of money they have altogether, thus, $(70 + 80 + 90) \div 2 = 120$. They have $120 altogether, so they have enough money.
- Students may say that if there is tax, they do not have enough money.

Lesson 5

Objective:
- Determine whether a given number makes an inequality true.

1. Introduction

Use a balance scale and cubes to recreate the picture of the inequality on page 44. Ask students what other values x could have so the side with the variable (x) is heavier than the side with the number (2).

Give pairs or groups of students balance scales and cubes (from Lesson 2), and ask them to make other inequalities.

Ask students to share the inequalities they made and discuss some of the possible answers to their inequalities.

Discuss the top of page 44 and REMARKS.

2. Development

Have students study Example 15 and discuss the solutions.

Notes:
- The solution to an inequality is the set of values that makes the inequality true.
  5 is one value that will make the inequality true in (a) and (b), but it will not make the inequality true in (c).
- If necessary, remind students how to compare positive and negative numbers by drawing a number line on the board. Show how if $x = 5$, we can see from the number line that $5 > -10$, $5 > 4.9$ and $5 = 5$.

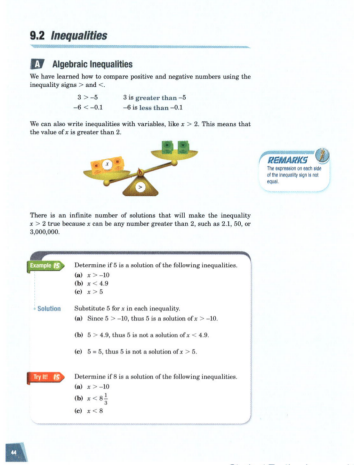

Student Textbook page 44

Have students do Try It! 15 and share and discuss their solutions.

3. **Application**

Have students study Example 16 and do Try It! 16.

Notes:
- In these problems, we are comparing the variable to a negative number. To illustrate the idea, draw a number line. In Example 16, mark y at -2 and then determine the relative positions of 0 and $-1\frac{1}{2}$.
- Encourage students who are confused to draw a number line.

Try It! 16 Answers

(a) Yes, $-4 > -4.01$

(b) No, $-4 < 3.9$

(c) No, $-4 = -4$

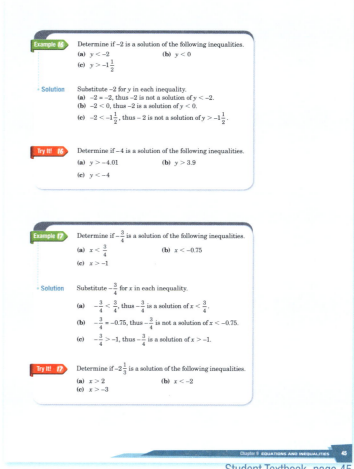

Student Textbook page 45

4. **Extension**

Have students study Example 17 and do Try It! 17.

Notes:
- Here, we are comparing the variable to a negative fraction. Draw a number line to remind students that $-\frac{3}{4}$ is in between 0 and -1.

Try It! 17 Answers

(a) No, $-2\frac{1}{3} < 2$

(b) Yes, $-2\frac{1}{3} < -2$

(c) Yes, $-2\frac{1}{3} > -3$

5. **Conclusion**

Summarize main points of the lesson.
- Unlike equations, the solution to an inequality has an infinite number of values.
- The expressions on each side of the inequality sign are not equal.
- We can use a number line to help us visualize an inequality.

★ **Workbook: Page 34**

Lesson 6

Objective:
- Graph inequalities using a number line.

1. Introduction

Draw a number line similar to the one on textbook page 46. Ask students:
- Where would the solutions to $x > -3$ be on the line? (To the right of -3.)
- Would -3 be part of the solution? (No, because all solutions must be greater than -3.)

Repeat with $x < 3$.

Introduce the symbol $\geq$ (greater than or equal to). Ask students where the solutions to $x \geq -3$ would be on the line, and how the solution is different from $x > -3$ (i.e., it includes -3).

Read and discuss the top of page 46 and the REMARKS.

Give students graph paper and rulers. Have them draw number lines to graph the solutions to $x > -3$, $x < -3$, $x \geq -3$, and $x \leq -3$. Make sure students understand that we use an open circle when we have > or <, and a closed circle when we have $\geq$ or $\leq$.

Notes:
- Remind students that on a horizontal number line, the numbers become greater going to the right and become less going to the left.
- The solution includes all the values indicated by the arrow, not just whole number values.

    Thus, the solution to $x < 5$ includes 4.3, $-\frac{1}{2}$, etc.

- When graphing the solutions to inequalities, make sure students draw the number lines directly on top of the horizontal lines on the graph paper, and use the vertical lines to draw the hash marks for the integers. Emphasize neatness and accuracy.

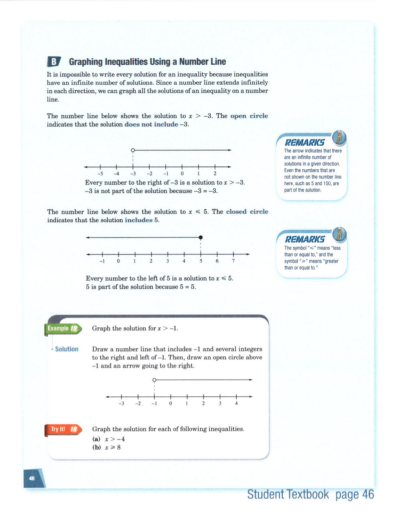

Student Textbook page 46

2. Development

Have students study Examples 18–20 and do Try It! 18–20 on their own, and then compare their solutions with partners or in a group.

Note:
- Make sure students are using the open and closed circles correctly, that the arrows are going in the proper direction, and that they are graphing the solutions neatly and accurately.

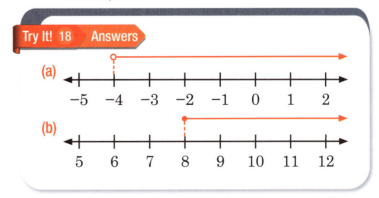

60

©2017 Singapore Math Inc. Dimensions Math® Teacher's Guide

Try It! 19   Answers

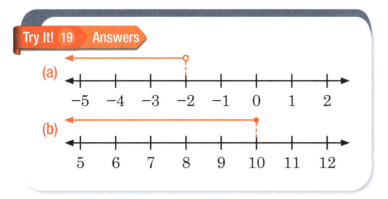

Example 20

- Students can use an arrow to indicate the approximate position of numbers that are not integers.

Try It! 20   Answers

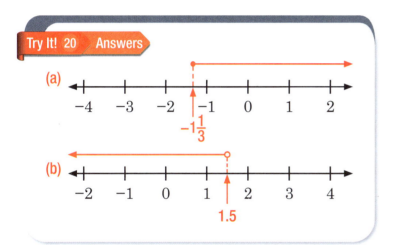

3. Application

Have students study Example 21 and do Try It! 21, then share and discuss their solutions.

Note:

- Discuss with students why the intervals of the number line do not have to always be 1. In Example 21, it makes more sense to use intervals of 100. In Try It! 21, students could use intervals of 5 or 10. Also, the graph does not always have to include 0.
- For Try It! 21, we are graphing the "usual" average temperature. Since it is "usually" below −10 °C, −10 is not part of the solution.

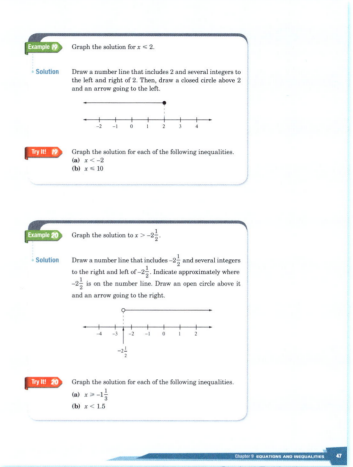

Student Textbook page 47

Try It! 21   Answers

(a) Let x °C represent the average temperature on the planet Mars. $x < -10$

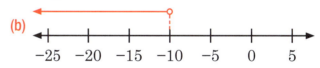

4. Extension

Have students study Example 22 and do Try It! 22. Have students share and discuss their solutions.

Notes:

- Encourage students to think about and choose the size of the increments on their number lines.

Notes:

Example 22 (a) and (b):
- Steven's position is "at most" 5 m above sea level, so 5 is part of the solution. For (b), Arlette must read for "a minimum" of 15 minutes, so 15 is part of the solution. Try It! 22 (a) and (b) are similar.
- There are situations where an open circle would be used. E.g., if the height of Steven's position was less than 5 m above sea level.
- These situations involve boundaries, because Steven's height and the length of time Arlette can read have limits. But since we do not know what that limit is, we draw lines with arrows as if there were no limit. Try It! 22 (a) and (b) are similar.

Example 22 (c):
- This is the first time students have seen an interval represented by a system of inequalities, so it is important to discuss this with students. Help them see that there are two conditions, $t \geq 0$ and $t \leq 30$. Thus, $0 \leq t \leq 30$ means that t is greater than or equal to 0, but it's less than or equal to 30 (see DISCUSS).
- Unlike problems (a) and (b), in (c) there are two boundaries, 0 and 30. Kini cannot read less than 0 minutes or more than 30 minutes so the solution includes 0, 30, and everything in between.
- Try It! 22 (c) is a similar situation. The speed Aisha can run includes 0 km/h and 8 km/h, and every speed in between.
- There are situations where open circles would be used (e.g., if Kini reads more than 0 minutes and less than 30 minutes).

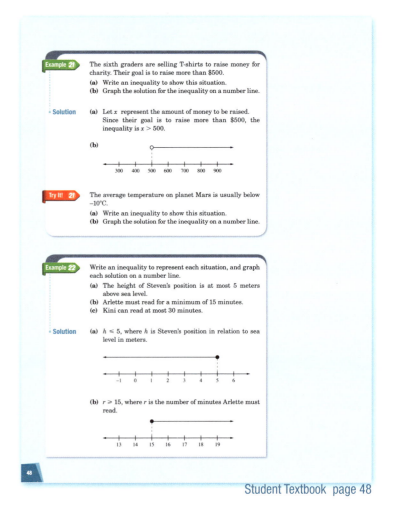

Student Textbook page 48

Try It! 22   Answers

(a) Let $x represent the money Jeremy has in his wallet. $x \geq 12$

(b) Let x represent the number of stamps Tom has in his album. $0 \leq x \leq 50$

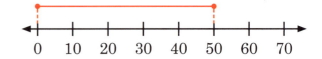

(c) Let x represent the speed Aisha can run. $0 \leq x \leq 8$

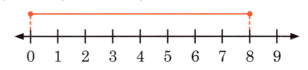

Lesson 7

5. Conclusion

Summarize main points of the lesson.
- We can use a number line to show the solutions to inequalities graphically. A closed circle means that value is included, and an open circle means that value is not included. An arrow indicates that there are an infinite number of solutions in the direction of the arrow.
- In many real-life situations, there are lower and upper limits to the solution. We can graph these situations using points connected by a line.

★ **Workbook: Page 35**

> **Objective:**
> - Consolidate and extend the material covered thus far.

Have students work with a partner or in groups. Students should try to solve the problems by themselves first, then compare solutions with their partner or group. If they are confused, they can discuss together.

Observe students carefully as they work on the problems. Give help as needed individually or in small groups.

BASIC PRACTICE

1. (a) Yes (b) Yes
   (c) No

2. (a) Yes (b) No
   (c) Yes

3. (a)

   (b)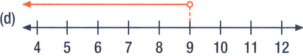

FURTHER PRACTICE

4. (a) Yes (b) No
   (c) Yes

5. (a) Yes (b) No
   (c) Yes

6. (a)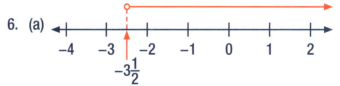

   (b)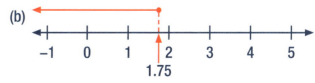

 **MATH@WORK**

7. $12x = 72$
   $x = 6$
   or, $72 \div 12 = 6$
   He needs to work at least 6 hours.

8. $60x = 2{,}400$
   $x = 40$
   or, $2{,}400 \div 60 = 40$
   She has to save at least 40 weeks.

9. $4x = 80$
   $x = 20$
   or, $\$80 \div 4 = \$20$
   $20 is the most he can spend on a pair of pants so the inequality is $x \leq 20$.

Note on Problem #9: Students may write $4x \leq 80$, which is the solution given in the back of the book. However, students have not learned how to solve inequalities yet, so they should solve the problem using algebra or arithmetic. Then, they should think about how to write an inequality and draw a graph to represent the answer.

 **BRAIN WORKS**

10. $6 \leq t < 20$

Note on Problem #10: This problem involves two different inequality signs. Mandy has to work at least 6 hours, so 6 is part of the solution. She has to work less than 20 hours, so 20 is not a part of the solution.

11. $5 \leq m < 10$

Note on Problem 11: "She has never milked less than 5 gallons" means it could be exactly 5 gallons or more, so 5 is part of the solution. She never milks 10 gallons, so 10 is not part of the solution.

Student Textbook page 50

12. (a) $\$(80 + 10h)$

    (b) Students may write $80 + 10h \leq 120$ which is the answer in the back of the book. To graph this, however, they need to first determine the greatest number of hours the member can use the badminton hall:

    $80 + 10h = 120$
    $10h = 40$
    $h = 4$

    The greatest number of hours a member can use the club is 4. The solution can be written and graphed as $h \leq 4$.

    (c) 4 hours

Lesson 8

13. (a) $6x = 30$
    $x = 5$
    or, $30 \div 6 = 5$

    Since the number of cakes is more than 30, Rachel has to work more than 5 days so the inequality is $x > 5$.

    Note: On textbook page 249, the solution is given as $6x > 30$. Again, students have not learned to solve inequalities algebraically yet, so they should use algebra or arithmetic to determine that she needs to work more than 5 days, then write and graph the solution.

    (b) She has to work more than 5 days, so 5 is not part of the solution. At most she can work 8 days, so 8 is part of the solution. $5 < x \leq 8$

14. (a) $250 \div 2 = 125$, so to make $250 they need to wrap 125 presents. Since they need to make more than $250, they need to wrap at least 126 presents.
    (b) $250 \div 1.5 = 166.666\ldots$ which is a little less than 167, so they need to wrap at least 167 presents.

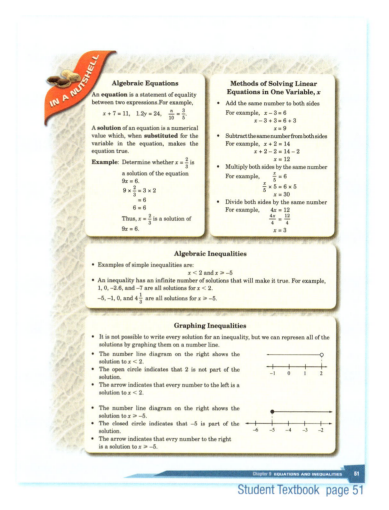

Student Textbook page 51

Objectives:

- Summarize and reflect on important ideas learned in this chapter.
- Use a spreadsheet to evaluate expressions and solve equations.

This lesson could be done in class or assigned to students to do independently at home or at school.

1. **In a Nutshell**

Use this page to summarize the important ideas learned in this chapter. Give examples as needed.

2. Write in Your Journal

Have students do the writing activity and share their answers. Answers will vary.

3. Extend Your Learning Curve

This activity can be completed in class or done as an independent assignment.

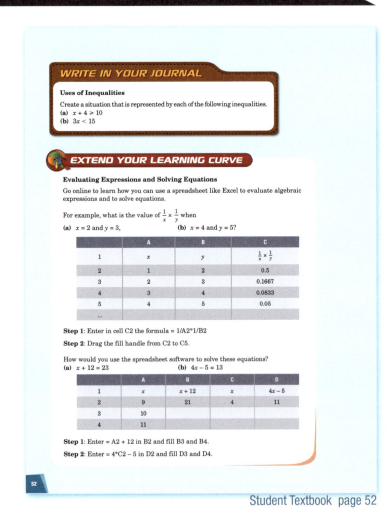

Student Textbook page 52

Lesson 9

Objective:

- Apply the model method to solve complex word problems involving algebraic equations.

1. Problem Solving

Have students study Examples 1–3 and do Try It! 1–3. Students can work individually or in pairs or groups. If students are working in pairs or groups they should try the problems independently first and then share and discuss with their groups.

Have students share and discuss their solutions and methods.

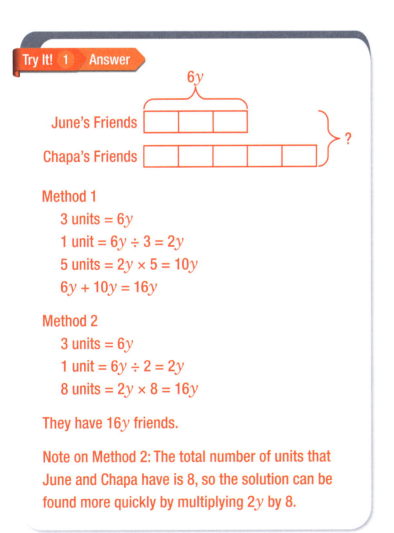

Try It! 1 Answer

Method 1
- 3 units = $6y$
- 1 unit = $6y \div 3 = 2y$
- 5 units = $2y \times 5 = 10y$
- $6y + 10y = 16y$

Method 2
- 3 units = $6y$
- 1 unit = $6y \div 2 = 2y$
- 8 units = $2y \times 8 = 16y$

They have $16y$ friends.

Note on Method 2: The total number of units that June and Chapa have is 8, so the solution can be found more quickly by multiplying $2y$ by 8.

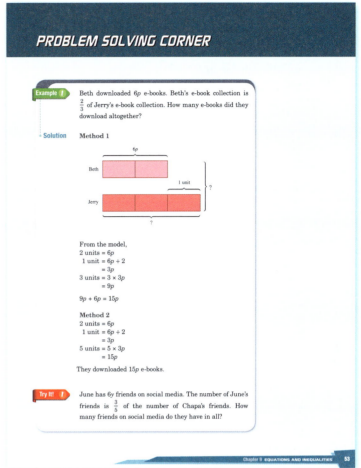

Student Textbook page 53

Note on Example 2: This can also be written without the model as:

Jon's age 4 years ago $\to k$
Jon's age now $\to k + 4$
Mother's age now $\to 3(k + 4) = 3k + 12$
Mother's age in 5 years $\to (3k + 12) + 5 = 3k + 17$

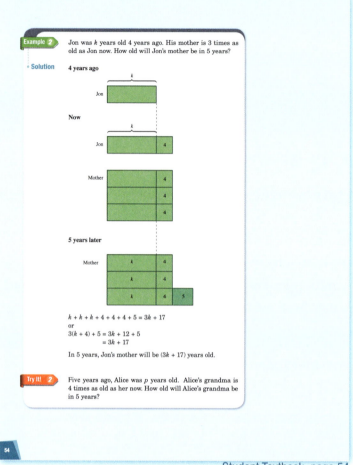

Student Textbook page 54

Try It! 2   Answer

5 years ago
  Alice | p |

Now
  Alice | p | 5 |
  Grandma | p | 5 |
p	5
p	5
p	5

5 year later
  Grandma | p | 5 |
p	5
p	5
p	5	5

$p + p + p + p + 5 + 5 + 5 + 5 + 5 = 4p + 25$
(or, $4 \times p + 5 \times 5 = 4p + 25$)

Alternatively,
  Alice's age 5 years ago $\to p$
  Alice's age now $\to p + 5$
  Grandma's age now $\to 4(p + 5) = 4p + 20$
Grandma's age in 5 years $\to$
  $(4p + 20) + 5 = 4p + 25$
In 5 years, Alice's grandma will be $(4p + 25)$ years old.

68

©2017 Singapore Math Inc. Dimensions Math® Teacher's Guide 6

Try It! 3   Answers

(a) $x° + 110° + 58° = 180°, x = 12$

(b) $27° + x° + 30° = 90°, x = 33°$

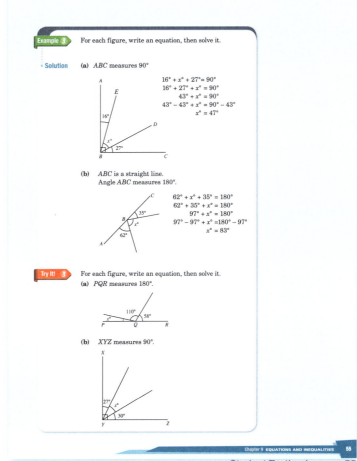

Student Textbook page 55

2. Practice

Have students do Practice Questions 1 – 10. Students can work together and help each other, but they should try the problems independently first.

Answers

1. Binh's age now → $k - 3$
    Father's age now → $4(k - 3) = 4k - 12$
    Father's age 5 years ago →
    $(4k - 12) - 5 = 4k - 17$
    His father was $(4k - 17)$ years old.

2. p pens → 120 g
    1 pen → $\frac{120}{p}$ g
    70 pens → $70 \times \frac{120}{p}$ g = $\frac{8{,}400}{p}$ g

3. PQS, SQR — 90°

    PQR → 3 units = 90°
    SQR → 1 unit = 90° ÷ 3 = 30°
    PQS → 2 units = 30° × 2 = 60°

4. $180 - (42 + 42) = 96$; 96°

5. Angle 1, Angle 2 — 180°

    Both angles → 6 units = 180°
    Angle 2 → 1 unit = 180° ÷ 6 = 30°
    Angle 1 → 5 units = 30° × 5 = 150°

6. (a) 12 cans (weight of soup and cans together)
    → w grams
    1 can → $\frac{w}{12}$ g
    Weight of soup in 1 can → $\left(\frac{w}{12} - 110\right)$ g
    (b) $\left(\frac{8{,}400}{12} - 110\right)$ g = 590 g

7. (a) Mala → d
    Jeff → $d + 11$
    Hunter → $2d - 20$
    $d + (d + 11) + (2d - 20) = 4d - 9$
    They made $(4d - 9)$ donuts.
    (b) $4d - 9 = 123$
    $4d = 132$
    $d = 33$
    Jeff made $(d + 11)$ donuts.
    $33 + 11 = 44$
    Jeff made 44 donuts.

PRACTICE QUESTIONS

1. Binh will be k years old in 3 years. His father is 4 times as old as him. How old was Binh's father 5 years ago?

2. If p pens weigh 120 grams, how much will 70 pens weigh?

3. PQR measures 90°. It has been split into two angles, PQS and SQR. The measure of the two angles are in a ratio of 2 : 1. What are the measures of each angle?

4. The figure below shows a ray of light being reflected off a mirror. What is the measure of angle x?

5. The measures of two angles have a sum of 180°. The measures of the angles are in a ratio of 5 : 1. Determine the measures of the two angles.

6. A dozen cans of chicken soup, each of equal weight, weigh w grams in total. Each empty can weighs 110 grams.
   (a) What is the weight, in terms of p, of the chicken soup in each can?
   (b) If $w = 8{,}400$, find the weight of the chicken soup in each can.

7. Mala made d donuts, Jeff made 11 donuts more than Mala, and Hunter made 20 less than twice the number of donuts as Mala.
   (a) How many donuts did they make in total in terms of d?
   (b) If the three of them made a total of 123 donuts, how many donuts did Jeff make?

8. Aaron, Jamal, and Cathy each received y candies from Ms. Li. Aaron gave away $\frac{1}{2}$ of his share. Jamal ate $\frac{1}{4}$ of his share. Cathy ate all her candies. How many candies did they have left altogether?

9. Mr. Smith bought three identical cell phones and two identical chargers. One cell phone costs $470 more than one charger. If the total bill was $1560, and there was no tax, how much did one charger cost?

10. A student claims that these two operations are the same:

A number x is multiplied by 3 and then added by 4.	A number x is added to 4 and then multiplied by 3.

    Explain whether his claim is correct.

Student Textbook page 56

8. All the candies $\longrightarrow y + y + y = 3y$

Candies Aaron has left $\longrightarrow y - \frac{1}{2}y = \frac{1}{2}y$

Candies Jamal has left $\longrightarrow y - \frac{1}{4}y = \frac{3}{4}y$

Candies Cathy has left $\longrightarrow y - y = 0$

Candies they all had left $\longrightarrow \frac{1}{2}y + \frac{3}{4}y + 0 = \frac{5}{4}y$

They had $\frac{5}{4}y$ candies left.

9. One charger

c

One phone

c	$470

Two chargers

c
c

Three phones

c	$470
c	$470
c	$470

$\}$ $1,560

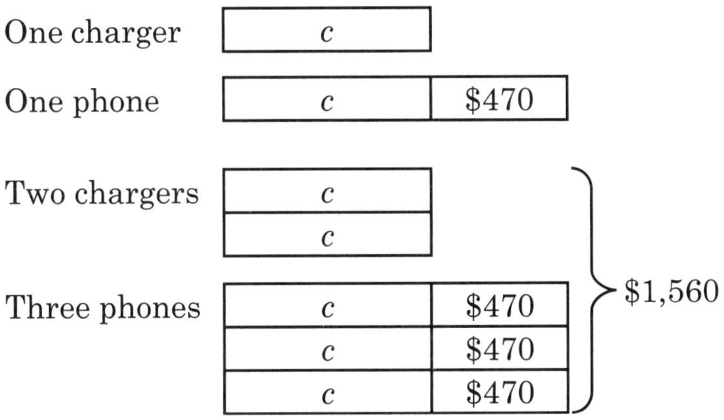

Cost of 2 chargers and

3 phones $\longrightarrow$ $1,560

Cost of 5 chargers $\longrightarrow$ $1,560 - (3 \times $470) = $150

Cost of 1 charger $\longrightarrow$ $150 \div 5 = $30

10. No, $3x + 4 \neq 3(x + 4)$

Notes

Chapter 10: Coordinates and Graphs

Lesson	Objectives	Class Periods	Textbook & Workbook	Teacher's Guide Page	Additional Materials Needed
1	• Use ordered pairs of coordinates to plot points.	1	TB: 58–64	79	
2	• Draw polygons on the coordinate plane.	1	TB: 65–67 WB: 40–43	87	Graph paper, rulers
3	• Consolidate and extend the material covered thus far.	1	TB: 68–69	90	Graph paper, rulers
4	• Determine whether a line joining two points is vertical or horizontal based on the coordinates of two points. • Find the distance between two points on a horizontal or vertical line on the coordinate plane. • Determine whether two points are located in the same quadrant given their coordinates.	2	TB: 70–78	94	Graph paper, rulers
5	• Find the distance between two points that are in different quadrants. • Apply this to find the area and perimeters of rectangles whose points are in different quadrants. • Given two points, determine the coordinates of two other points of a quadrilateral.	1	TB: 77–83 WB: 44–56	103	Graph paper, rulers
6	• Consolidate and extend the material covered thus far.	1	TB: 84–86	108	Graph paper, rulers
7	• Identify independent and dependent variables and describe their relationships.	1	TB: 86–88 WB: 57	111	

Continues on next page.

Chapter 10: Coordinates and Graphs

Lesson	Objectives	Class Periods	Textbook & Workbook	Teacher's Guide Page	Additional Materials Needed
8	• Represent relationships between variables using a table, and write an equation to represent this relationship.	1	TB: 89–94 WB: 58–60	113	Pattern block triangles (1-in side equilateral triangles), or toothpicks
9	• Graph equations and solve problems involving changes in quantities.	1	TB: 94–101 WB: 61–71	119	Graph paper, rulers, regular pencils, and colored pencils
10	• Consolidate and extend the material covered thus far.	1	TB: 102–105	126	Graph paper, rulers, regular pencils, and colored pencils
11	• Summarize and reflect on important ideas learned in this chapter, and solve a non-routine problem.	1	TB: 106–107	132	

Students used horizontal and vertical number lines to locate points in one dimension when they learned about positive and negative numbers in Chapter 4: Negative Numbers. They also learned that the absolute value of a number is the distance of that number from 0. In this chapter, students will build on these concepts to show the positions of points in two-dimensional space on a coordinate plane.

A **plane** is a flat two-dimensional surface that extends infinitely in all directions. A **coordinate plane** consists of a horizontal and a vertical line that intersect at right angles to divide the plane into four sections, called **quadrants**. The horizontal line is called the x-axis and the vertical line is called the y-axis. The lines intersect at the **origin** (0, 0), which is the location of 0 on each number line.

We can find any point on the graph by naming the coordinates of the point, which is an ordered pair of numbers. The first number in the pair is the x-coordinate and it indicates the location of the point along the x-axis. The second number in the pair is the y-coordinate and it indicates the location of the point along the y-axis. We can determine what quadrant the point is in based on the signs of the coordinates. In the first quadrant, both coordinates are positive (+, +). Moving counter-clockwise from the first quadrant are the second quadrant (−, +), the third quadrant (−, −), and the fourth quadrant (+, −) respectively.

Chapter 10: Coordinates and Graphs

For example, the coordinates of point Q (4.5, −3) locate it 4.5 units from 0 in the positive direction along the x-axis and 3 units from 0 in the negative direction along the y-axis, which puts it in the 4th quadrant. Coordinate pairs are shown in the form of (x, y), where x indicates the location of the point relative to the x-axis and y indicates its location relative to the y-axis.

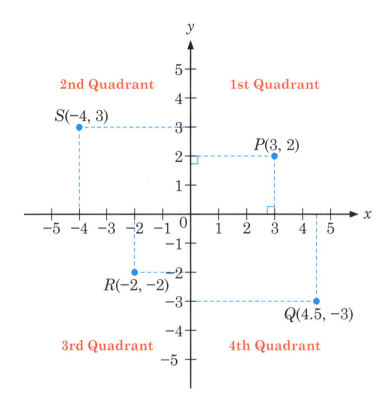

The absolute values of each of the coordinates tell us the distance of the point from the x and y axes, respectively. For point S (−4, 3), if we draw a vertical line from the point downwards, it will cross the x-axis at −4. This means that the point is $|-4| = 4$ units from the y-axis. If we draw a horizontal line from the point to the right, it will cross the y-axis at 3. Thus, the point is 3 units from the x-axis.

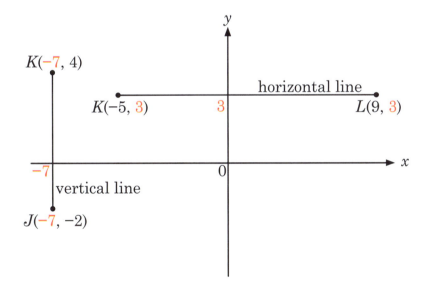

We can determine whether two points form a horizontal or a vertical line when connected by looking at their coordinates.

If the distances of the points from the x-axis are the same (that is, their y-coordinates are the same), then a line drawn between the two points will be horizontal (parallel to the x-axis). If the points are in different quadrants, the line will cross the y-axis at the value of the y-coordinates. If the distances of the points from the y-axis are the same (that is, their x-coordinates are the same), then a line drawn between the two points will be vertical (parallel to the y-axis). If the points are in different quadrants, the line will cross the x-axis at the value of the x-coordinates.

We can also use these ideas to find the distance between the two points.

If two points on a horizontal or vertical line are in the same quadrant, we can find the distance between the points by finding the difference between the absolute values of the coordinates that are different.

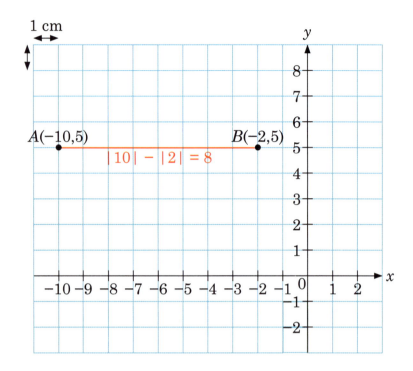

A and B are in the same quadrant. Since they are equidistant from the y-axis, we need only concern ourselves with describing their distance from each other with respect to the x-axis. The absolute values of the y-coordinates tell us how far the points are from the x-axis. Thus, to find the distance of the points from each other, we subtract the absolute values of the x-coordinates.

If two points on a horizontal or vertical line are in different quadrants, we can find the distance between the points by finding the sum of the absolute values of the coordinates that are different.

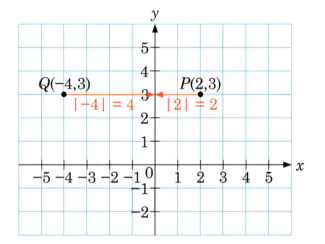

Q and P are in different quadrants. Since they are equidistant from the x-axis, we need only concern ourselves with describing their distance from each other with respect to the y-axis. The absolute values of the x-coordinates tell us how far the points are from the y-axis. Thus, to find the distance of the points from each other, we add the absolute values of the x-coordinates.

$|-4| + |2| = 4 + 2 = 6$

The distance is 6 cm.

We can also use the coordinate plane to model the relationship between two quantities that are changing together.

The table below shows the relationship between time, t hours, and distance, d miles, for a car traveling at a constant speed of 40 mph.

t (hours)	0	1	2	3	4
d (miles)	0	40	80	120	160

Students saw these types of quantitative relationships when they learned rate in Chapter 6. They also learned rate formulas such as **Distance = Speed × Time**. The equation $d = 40t$ will give us the distance given any number of hours. The total distance in miles depends on the number of hours the car travels. Thus, we call t the independent variable and we call d the dependent variable.

The values of t and d can be plotted as ordered pairs on a coordinate graph.

By convention, the independent variable is usually graphed on the x-axis and the dependent variable on the y-axis. The straight line connecting the points signifies that all of the values on the line are part of the solution (e.g., in 1.5 hours, the car travels 60 miles).

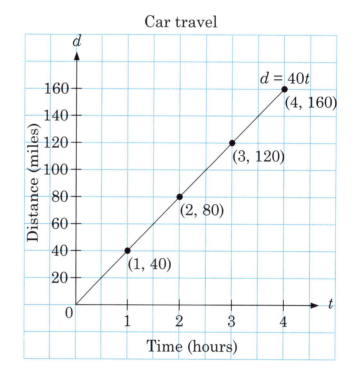

Notes

Lesson 1

Objective: Use ordered pairs of coordinates to plot points.

1. Introduction

Discuss page 57 and state the goals of the chapter, LET'S LEARN TO …

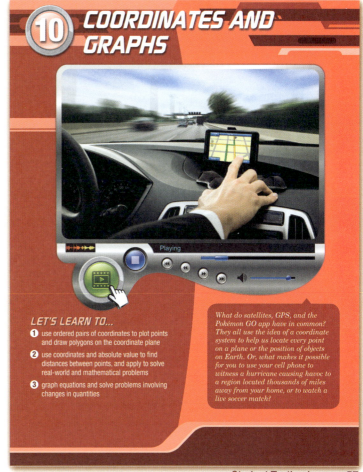

Student Textbook page 57

Read the story about Descartes (MATHS MATTER), and ask students how they would locate the position of the fly on the ceiling.

Read and discuss pages 58 – 59.

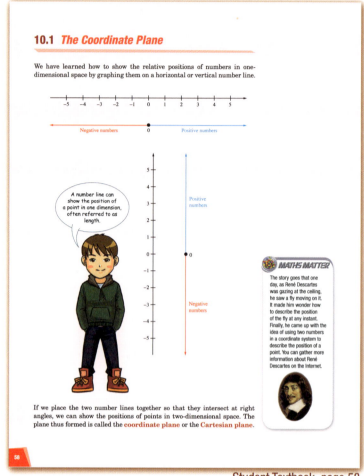

Student Textbook page 58

Give students graph paper and rulers, and have them draw the coordinate axes similar to page 59. (See the graph in Example 2 on page 62 for an example of how to draw a coordinate graph.)

Note:
- Make sure students use a ruler and a sharp pencil, and that they draw the axes directly over the lines on the graph paper, not in between. Label the origin, and, using the graph paper intersections, label the scale of each axis with positive and negative integers.

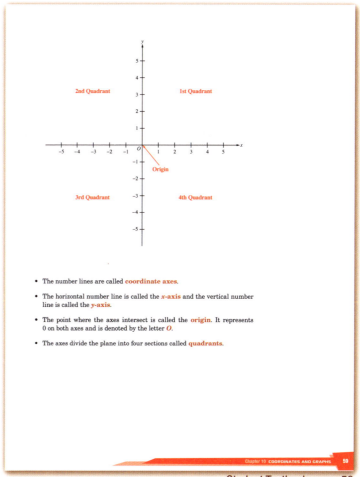

Student Textbook page 59

2. Development

Read and discuss the top of page 60.

Have students plot the points *P*, *Q*, *R*, and *S* on their own graph paper as shown on page 60.

Read and discuss the summary on the bottom of page 60.

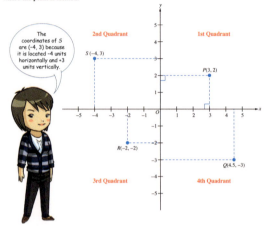

Student Textbook page 60

3. Application

Have students study Examples 1 – 2 and do Try It! 1 – 2.

Notes:
- For Example 1 (b), help students see that the x-coordinate (-7.5) indicates that the point is between 7 and 8 units away from the y-axis, as measured by the x-axis.
- For Example 1 (d), help students see that the y-coordinate (-2.5) is between -2 and -3 units away from the x-axis as measured by the y-axis.

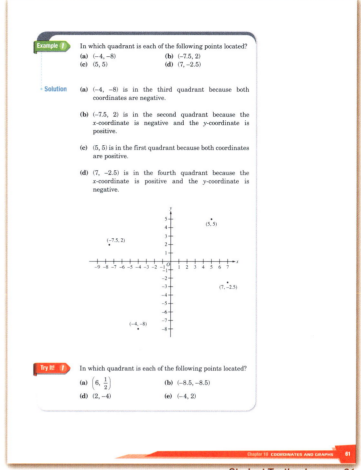

Student Textbook page 61

Discuss the red arrows on the solution of Example 2, and help students see that the coordinates tell the distance of the point from the x- and y-axes.

- The x-coordinate describes the distance of the point from the y-axis. The y-coordinate describes the distance of the point from the x-axis.

Discuss point E and what the character is saying.

- If the y-coordinate is 0, the point will be on the x-axis, because it is 0 units away from the x-axis. It is also important to talk about ordered pairs such as (0, 2), which lie on the y-axis, and (0, 0) — the origin, which lies at the intersection of the two axes.

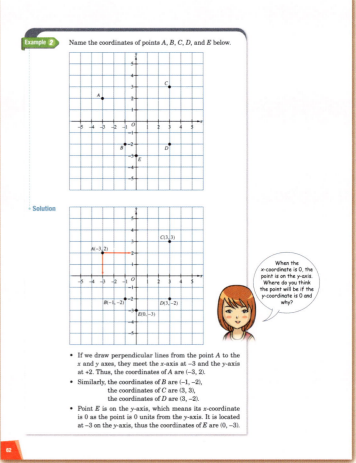

Student Textbook page 62

Try It! 2   Answers

$P(4, 1)$
$Q(-5, 0)$
$R(2, -4)$
$S(-4, -2)$
$T(-2, 3)$
$U(0, 5)$

4. Extension

Have students read and discuss Example 3 and do Try It! 3.

Note:
- Help students see that one unit on the scale on the axes is two units on the graph paper, because the points being plotted include numbers like 2.5 and −2.5 (i.e., the side of each square on the graph paper is 0.5 units). It is important for students to think about the numbers involved in the coordinates before drawing the graph, to decide how to draw the scales on the axes.

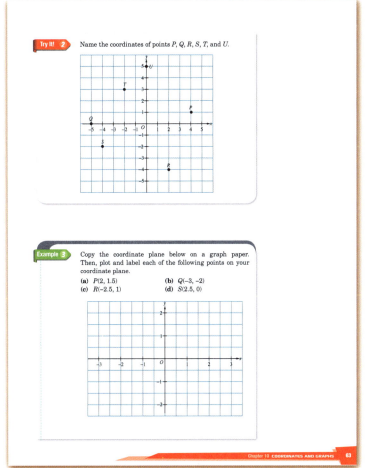

Student Textbook page 63

Try It! 3   Answers

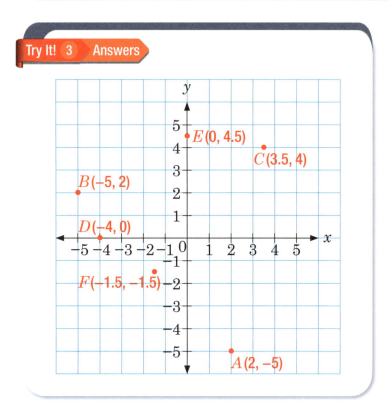

5. Conclusion

Summarize the important points of the lesson.
- We can use a coordinate plane to locate and plot points in two-dimensional space. The signs of the coordinates determine the quadrant the point is in.
- The absolute value of the x-coordinate tells the distance of the point from the y-axis, and the absolute value of the y-coordinate tells the distance of the point is from the x-axis.
- If the x-coordinate is 0, the point is directly on the y-axis. If the y-coordinate is 0, the point is directly on the x-axis. If both coordinates are 0, the point is at the origin.

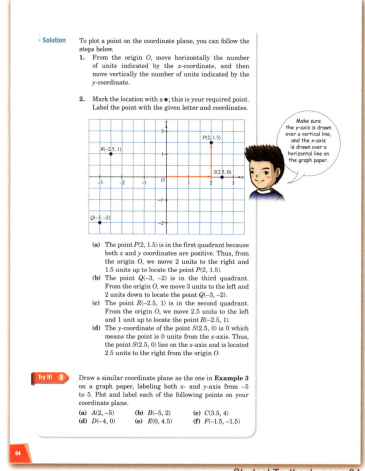

Student Textbook page 64

Lesson 2

Objective: Draw polygons on the coordinate plane.

1. Introduction

Give students graph paper and rulers, and have them do problems 1 – 3 of Class Activity 1.

2. Development

Have students work with a partner to do problems 4 – 6 of Class Activity 1.

Have some students share the shapes they drew with the class.

Use the bottom of page 66 to summarize the activity.

Note:

- Although the scales for each axis do not have to be the same when plotting points, the scales should be the same when drawing a polygon; otherwise, the figure will be skewed (distorted). Discuss this with the students.

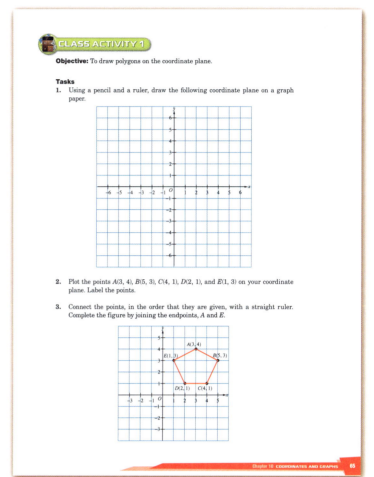

Student Textbook page 65

3. Application

Have students study Example 4 and do Try It! 4.

Notes:
- For Try It! 4, ask students to discuss how they chose their scales.
- For Try It! 4 (a), the x-coordinates and y-coordinates are multiples of 5, so a scale of 5 units is appropriate.
- For Try It! 4 (b), it makes more sense to have a scale of 1 on each axis.

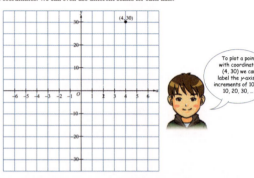

Student Textbook page 66

Try It! 4 Answers

(a) Isosceles triangle

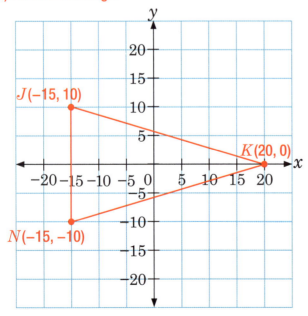

(b) Parallelogram

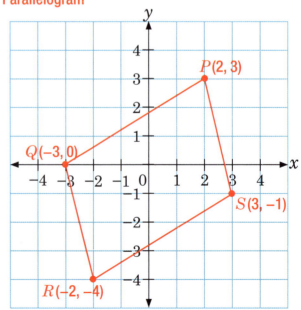

4. Extension

Refer students to the graph of Example 4. Ask them to compare the *x*- and *y*-coordinates of the points and discuss how they affect the orientation of the lines drawn for the sides of the polygon relative to the axes.

Notes:
- The adjacent sides of a rectangle form perpendicular lines. In this case, *AB* and *DC* are parallel to the *x*-axis, and *AD* and *BC* are parallel to the *y*-axis.
- In points *A* and *B*, the *y*-coordinates are the same. When joined, they form a horizontal line that intersects the *y*-axis at $y = 4$. This is also true for points *C* and *D*, which intersect the *y*-axis at $y = -3$.
- In points *A* and *D*, the *x*-coordinates are the same. When joined, they form a vertical line that intersects the *x*-axis at $x = -3$. This is also true for points *B* and *C*, which intersect the *x*-axis at $x = 1$.
- In the first printing of the textbook, there is an error in the coordinate plane diagram.

Have students draw a coordinate plane and plot the points $A(-3, 2)$ and $B(4, 2)$.
- Ask, "We want to plot two more points, *C* and *D*, to draw a rectangle. What are some possible coordinates for *C* and *D*?"
- Have students plot points *C* and *D* and draw rectangles. Then have them share and discuss the different rectangles they drew and the coordinates of *C* and *D*.
- Points *C* and *D* would have to have the same *x*-coordinates as *A* and *B*. There can be many different *y*-coordinates but both have to have the same value, e.g., $C(-3, 5)$, $D(4, 5)$.

Have students draw a coordinate plane and plot the points $W(2, 4)$ and $X(2, -2)$.
- Ask students to plot two more points, *X* and *Y*, and draw a rectangle, then share and discuss.
- Points *Y* and *Z* would have to have the same

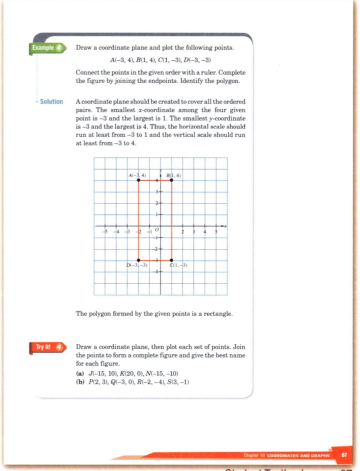

Student Textbook page 67

y-coordinates as *W* and *X*. There can be many different *x*-coordinates but both would have to have the same value, e.g., $Y(-1, -2)$, $Z(-1, 4)$.

5. Conclusion

Summarize the main points of the lesson.
- We can plot points on the coordinate plane and connect them with line segments to draw polygons and other shapes.
- When we draw a coordinate graph, we should look at the values of the *x*- and *y*-coordinates to determine the scale. When plotting points for polygons, both axes should use the same scale.
- Vertical lines drawn between two points will cross the *x*-axis at the same point. Horizontal lines drawn between two points will cross the *y*-axis at the same point.

★ **Workbook: Page 40**

Lesson 3

Objective: Consolidate and extend the material covered thus far.

Have students work together with a partner or in groups. Students should try to solve the problems by themselves first, then compare solutions with their partner or group. If they are confused, they can discuss together.

Observe students carefully as they work on the problems. Give help as needed individually or in small groups.

BASIC PRACTICE

1. (a) 2 (b) –6
   (c) 0 (d) –3.6

2. (a) –7 (b) 5.5
   (c) 0 (d) $\frac{3}{4}$

3. (a) Second quadrant (b) Fourth quadrant
   (c) First quadrant (d) Third quadrant
   (e) Second quadrant (f) Fourth quadrant

4. $A(4, -3)$, $B(0, -4)$, $C(-4, -5)$, $D(3, 4)$, $E(5, 0)$, $F(-2, 2.5)$

5. (a) (i) 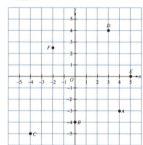 (ii) U
       (iii) T (iv)
       (v) □ (vi) ◆

   (b) $P(8, -35)$, $R(-5, -50)$, $S(5, 5)$

FURTHER PRACTICE

6. See coordinate plane to the right.
   (a) Second quadrant, $P(-3, 7)$, $W(-7, 1.5)$
   (b) $Q(7, -3)$
   (c) y-axis
   (d) x-axis
   (e) Origin

Student Textbook page 68

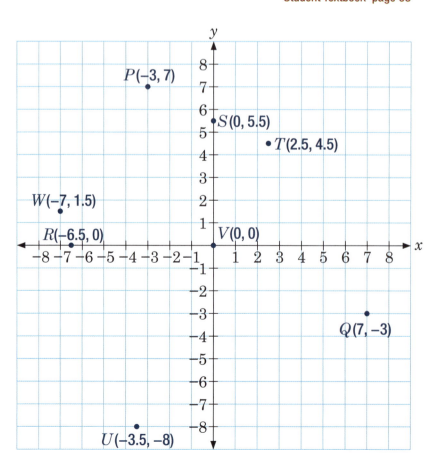

90

7. (a) Triangle
   (b) Rectangle or square

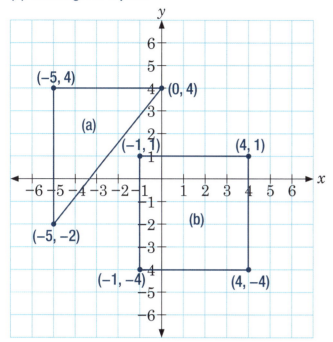

8. (a) (−3, 1)   (b) Bus stop
   (c) Cafe
   (d) 3 blocks due east and 3 blocks north
   (e) Yes

9. (a) Parallelogram   (b) Square
   (c) Trapezoid   (d) Pentagon
   (e) Hexagon   (f) Arrow
   Graphs for each of these question are found on the following two pages.

10. D will be the upper left vertex of the rectangle. The lower left vertex of the rectangle (A) has the coordinates (−10, 4) so the x-coordinate must be −10. The upper right vertex (C) has the coordinates (5, 12) so the y-coordinate must be 12. Thus, the coordinates of D are (-10, 12).

7. Draw a similar coordinate plane as the one in **Question 4** on a graph paper. For each question, plot and connect the points in the given order. Complete the shape by joining the end-points. Then give the best name for each shape.
   (a) (0, 4), (−5, −2), (−5, 4)
   (b) (4, 1), (−1, 1), (−1, −4), (4, −4)

8. Refer to the grid diagram to answer the questions below.
   Each square represents 1 block. You can only move horizontally or vertically. Do **not** move diagonally.

   (a) State the coordinates that defines Joan's location.
   (b) Which place is closer to Joan – the train station or the bus stop?
   (c) If Joan cycles 2 blocks due west and 4 blocks south, where would she reach?
   (d) Kelly is at the train station. Tell Kelly the shortest route to reach the bank.
   (e) If Joan from her original location and Kelly from the bank arrange to meet at the park, will this meeting place be equidistant for the girls? Explain your answer.

9. For each question, plot the points and connect them in the given order. Complete the figure by joining the endpoints. Identify the polygon.
   Use separate coordinate planes for each question.
   (a) (0, 6), (6, 4), (−1, −4), (−7, −2)
   (b) (1, 1), (−3, 5), (−7, 1), (−3, −3)
   (c) (4, 0), (0, 5), (−3, 4), (−5, −3)
   (d) (1, 2), (5, −2), (3, −6), (−1, −6), (−3, −2)
   (e) (0, 4), (2, 0), (0, −4), (−5, −3), (−7, 1), (−5, 5)
   (f) (0, 0), (2, −1), (2, 5), (−4, 2), (−2, 1), (−6, −3), (−4, −4)

10. You are graphing rectangle $ABCD$ in a coordinate plane. Three of the vertices of the rectangle are A(−10, 4), B(5, 4) and C(5, 12). Determine the coordinates of the point D.

11. You and your best friend, Michelle, plan to design some motifs. Michelle has planned her design using a set of points with the following coordinates.

    (5, 1), (3, −2), (3, −5), (0, −2), (−3, −5), (−3, −2), (−5, 1), (−2, 1), (0, 5), (2, 1)

    Plot the given points in a coordinate plane and connect them in the given order. What is Michelle's design?

    Similarly, plan your design using coordinates. You are to have at least 8 points for your design.

Student Textbook page 69

11. Her design is a star. (Example below.)
    Students will make different designs.

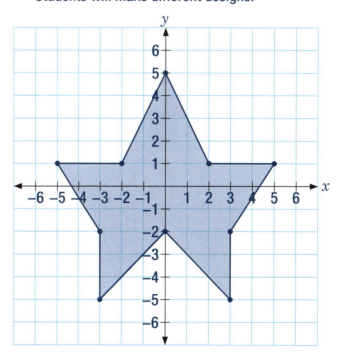

9. (a)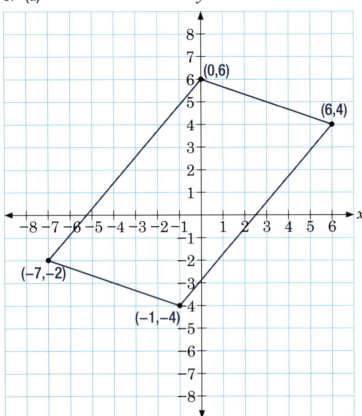

9. (b)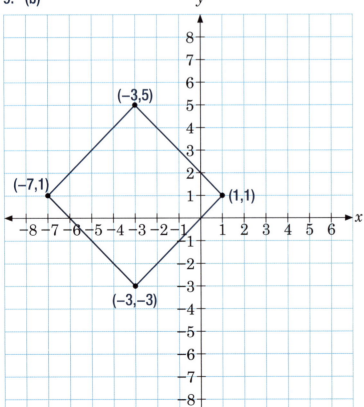

9. (c)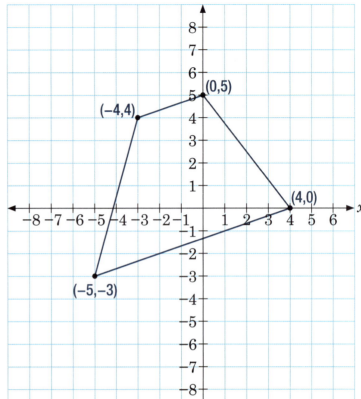

9. (d)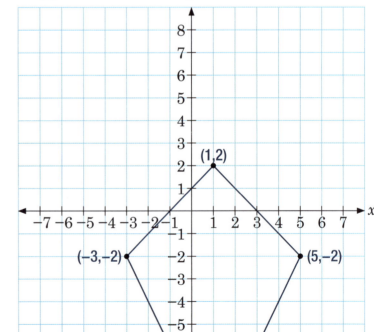

9. (e)

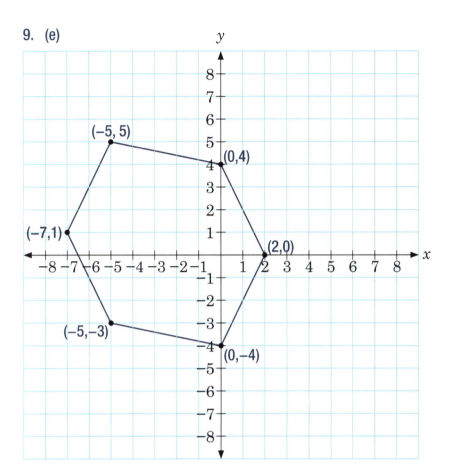

9. (f)

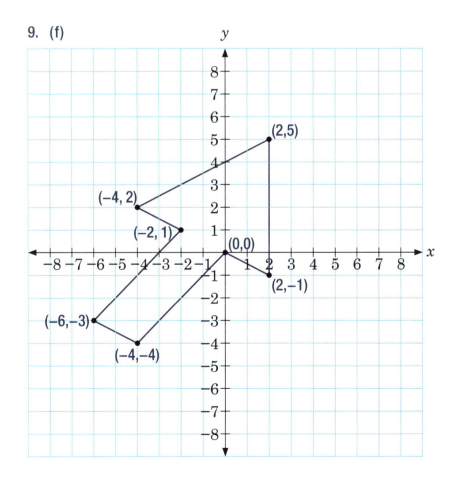

Lesson 4

Objectives:
- Determine whether a line joining two points is vertical or horizontal based on the coordinates of two points.
- Find the distance between two points on a horizontal or vertical line on the coordinate plane.
- Determine whether two points are located in the same quadrant given their coordinates.

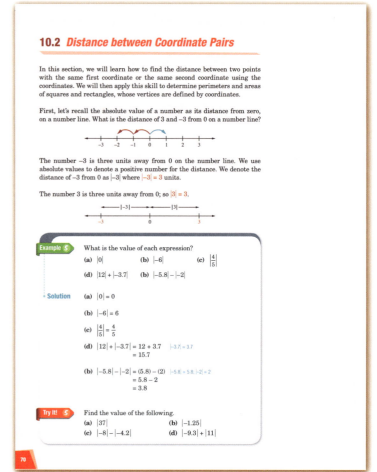

Student Textbook page 70

1. Introduction

Read and discuss the top of page 70 to review absolute value.

Have students study Example 5, do Try It! 5, and go over the solutions.

Notes:
- In **Dimensions Math® 6A** Chapter 4, students learned about absolute values but they did not add and subtract them. Help them see that they should find the absolute value first before adding or subtracting.
- In Example 5, (e) is mislabeled as (b) in the student textbook. Parentheses are used in the first step to clarify that we are subtracting $|-2|$, which is a positive number. However, it is not necessary to write the numbers in parentheses first.

Try It! 5 Answers

(a) $|37| = 37$

(b) $|-1.25| = 1.25$

(c) $|-8| - |4.2| = 8 - 4.2 = 3.8$

(d) $|-9.3| + |11| = 9.3 + 11 = 20.3$

2. Development

Have students do Class Activity 2 and discuss the results.

1.

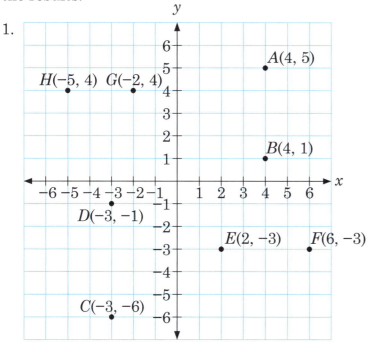

2.

Table 1

Points	Distance	Expression
$A(4, 5), B(4, 1)$	4	
$C(-3, -6), D(-3, -1)$	5	
$E(2, -3), F(6, -3)$	4	
$G(-2, 4), H(-5, 4)$	3	

(a) By counting the number of spaces (units) between the points on the graph paper.

(b) For A and B, subtract the y-values (5 – 1).
For C and D, subtract the absolute values of the y-values (6 – 1).
For E and F, subtract the x-values (6 – 2).
For G and H, subtract the absolute values of the x-values (5 – 2).

(c) For A and B, and C and D, the x-coordinates are the same. For E and F, and G and H, the y-coordinates are the same.
To find the distance between the points, find the difference between the absolute values of the coordinates that are not the same.

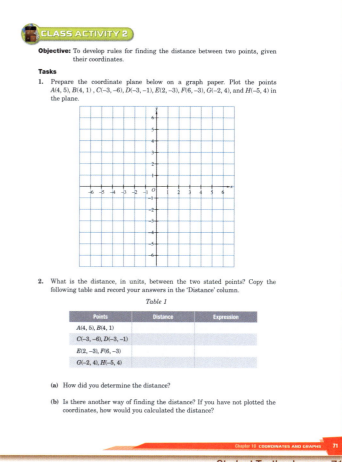

Student Textbook page 71

(d)

Table 1

Points	Distance	Expression
$A(4, 5), B(4, 1)$	4	$\|5\| - \|1\|$
$C(-3, -6), D(-3, -1)$	5	$\|-6\| - \|-1\|$
$E(2, -3), F(6, -3)$	4	$\|6\| - \|2\|$
$G(-2, 4), H(-5, 4)$	3	$\|-5\| - \|-2\|$

Note:
- For points C and D, students may write 6 – 1. For points G and H, students may write 5 – 2. Point out that these numbers come from finding the absolute values of the coordinates.

3.

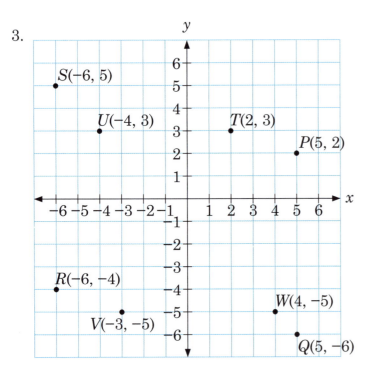

4. Table 2

Points	Distance	Expression
$P(5, 2), Q(5, -6)$	8	$|-6| + |2|$
$R(-6, -4), S(-6, 5)$	9	$|-4| + |5|$
$T(2, 3), U(-4, 3)$	6	$|2| + |-4|$
$V(-3, -5), W(4, -5,)$	7	$|-3| + |4|$

(b) For P and Q, and R and S, the x-coordinates are the same and the y-coordinates are different and have opposite signs.

For each set of points, the points are in different quadrants.

The distance can be calculated by adding the absolute values of the coordinates that are different.

When the signs of the coordinates are the same, subtract the smaller absolute value from the larger absolute value to find the distance between the points.

When the signs are different, add the absolute values.

When the signs are the same, it means the points are in the same quadrant. We find the difference between the absolute values to find the distance between the two points. When the signs are different, it means the points are in different quadrants. We find the sum of the absolute values.

Note:

- Point out to students that absolute values are added because the distance of each point to a particular axis must be determined before being combined to find the distance between the two points. For example, for points P and Q, the x-coordinates are the same. The y-coordinates are different. Point P is in the first quadrant and it is 2 units away from the x-axis. Point Q is in the fourth quadrant and it is 6 units away from the x-axis. Thus, the points are $(|2| + |-6|)$ units away from each other.

5. (a) They are the same. Vertical lines.
   (b) They are the same. Horizontal.

Summarize Class Activity 2 by reading and discussing page 73.

Notes:
- Help students understand that even if you move in a negative direction, the distance moved is always positive. (See what the character is saying on the right side of the page.) For example, a submarine that dives 100 m is moving in a negative direction from sea level, but the distance moved is still 100 m.
- Point out that these rules to find the distance between two points only apply if the points in question are on the same horizontal or vertical line. Students will learn how to find the distance between two points on the same diagonal line in **Dimensions Math® 8**.

3. Application

Have students study Examples 6–8 and do Try It! 6 – 8.

Student Textbook page 73

Note:

- For Try It! 6, have students use a regular pencil to draw the coordinate graph and plot the points, then use a colored pencil and ruler to connect the points with a line segment.

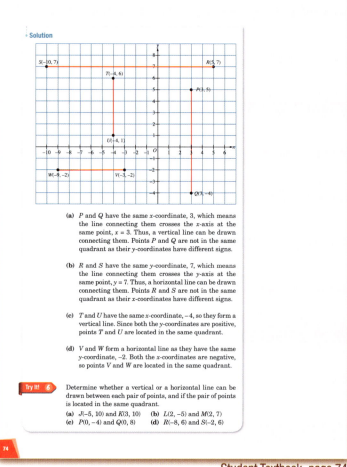

Student Textbook page 74

Try It! 6 Answers

(a) J and K have the same y-coordinate so the line segment crosses the y-axis at $y = 10$. It is a horizontal line. They are not in the same quadrant because the signs of the x-coordinates are different.

(b) L and M have the same x-coordinate so the line segment crosses the x-axis at $x = 2$. It is a vertical line. They are not in the same quadrant because the signs of the y-coordinates are different.

(c) P and Q have the same x-coordinate so the line segment crosses the x-axis at $x = 0$. It is a vertical line. They are not in any quadrant because the line segment is on the y-axis.

(d) R and S have the same y-coordinate so the line segment crosses the y-axis at $y = 6$. It is a horizontal line. They are in the same quadrant because the signs of the x-coordinates are both negative.

Notes:
- For Example 7 (a), remind struggling students of the signs of the coordinates in each quadrant. (See page 60.) If the x-coordinates and y-coordinates of two points have the same sign (positive or negative), the points are in the same quadrant.
- In Example 7 (b), the points are in the same quadrant and they form a horizontal line. Thus, we can find the distance between the points by finding the difference between the absolute values of the x-coordinates.

Try It! 7   Answers

(a) Both A and B have a negative x-coordinate and a positive y-coordinate, so they are both in the same quadrant (quadrant 2).
$|-8| - |-1| = 8 - 1 = 7$; 7 cm

(b) Both C and D have a negative x-coordinate and a negative y-coordinate, so they are both in the same quadrant (quadrant 3).
$|-12| - |-5| = 12 - 5 = 7$; 7 cm

(c) Both E and F have a positive x-coordinate and a negative y-coordinate so they are both in the same quadrant (quadrant 4).
$|20| - |3| = 20 - 3 = 17$; 17 cm

(d) Both G and H have a positive x-coordinate and a positive y-coordinate so they are both in the same quadrant (quadrant 1).
$|10| - |0.5| = 10 - 0.5 = 9.5$; 9.5 cm

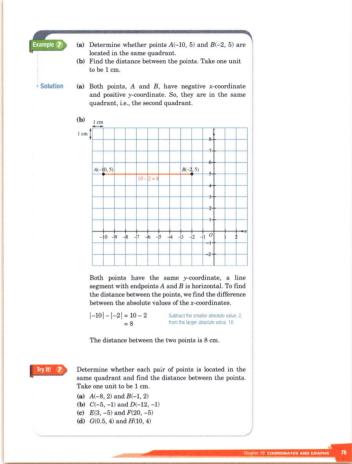

Student Textbook page 75

Notes:
- In Example 8 (a), since the x-coordinates are the same, we only need to see if the y-coordinates have the same sign to determine if they are in the same quadrant.
- In Example 8 (b), the points are in the same quadrant and they lie on a vertical line. Thus, we can find the distance between the points by finding the difference between the absolute values of the y-coordinates.

Try It! 8 Answers

(a) The y-coordinates have the same sign, so they are in the same quadrant.
$|-10| - |-2| = 10 - 2 = 8$; 8 cm

(b) The y-coordinates have the same sign, so they are in the same quadrant.
$|12| - |3| = 12 - 3 = 9$; 9 cm

(c) The y-coordinates have the same sign, so they are in the same quadrant.
$|-19| - |-4| = 19 - 4 = 15$; 15 cm

(d) The y-coordinates have the same sign, so they are in the same quadrant.
$|5| - |1.5| = 5 - 1.5 = 3.5$; 3.5 cm

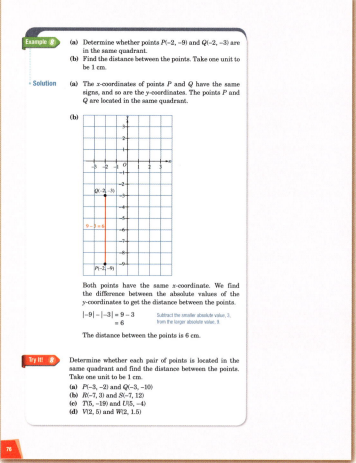

Student Textbook page 76

4. **Extension**

Have students study Example 9 and do Try It! 9.

Notes:
- Here, students are applying what they learned to find the area and perimeter of a rectangle. Remind students of the formulas for area and perimeter of rectangles. (See REMARKS.)
- Give students graph paper and have them draw the rectangle on a coordinate graph, then count the units between the points (Method 1).

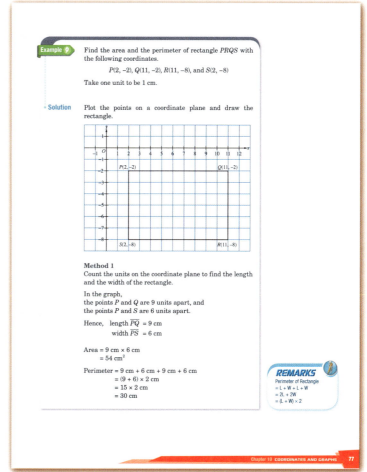

Student Textbook page 77

Note:
- Although Method 2 can be used without drawing the rectangle on a coordinate grid, students should make a quick sketch of the rectangle so they know which points form the horizontal side (the points with the same *y*-coordinate) and which points form the vertical side (the points with the same *x*-coordinate).

Try It! 9   Answers

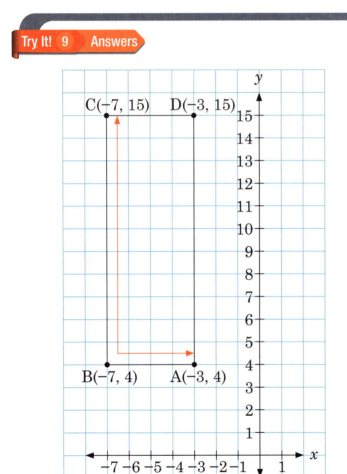

A and B are 4 units apart, $\overline{AB}$ = 4 cm
B and C are 11 units apart, $\overline{BC}$ = 11 cm
Area = 4 cm × 11 cm = 44 cm²
Perimeter = (4 + 11) × 2 cm = 30 cm

Method 2
Length $\overline{AB}$ = |–7| – |–3| = 7 – 3 = 4 cm
Width $\overline{BC}$ = |15| – |4| = 15 – 4 = 11 cm
Area = 4 cm × 11 cm = 44 cm²
Perimeter = (4 + 11) × 2 cm = 30 cm

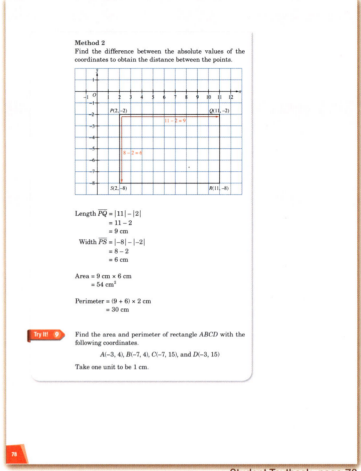

Student Textbook page 78

5. Conclusion

Summarize the main points of the lesson.
- We can determine whether two points form a vertical or horizontal line by looking at their coordinates. If they have the same *x*-coordinate, they form a vertical line. If they have the same *y*-coordinate, they form a horizontal line.
- We can determine the quadrant that two points are in by looking at the signs of the coordinates.
- When two points are in the same quadrant, we can calculate the distance between the points by finding the difference between the absolute values of the coordinates that are different.

Lesson 5

Objectives:
- Find the distance between two points that are in different quadrants.
- Apply this to find the area and perimeters of rectangles whose points are in different quadrants.
- Given two points, determine the coordinates of two other points of a quadrilateral.

1. Introduction

Give students graph paper and have them draw a coordinate graph. Have them plot points $P(2, 3)$ and $Q(-4, 3)$ in Example 10.

Discuss the two methods. Have students use a colored pencil to draw arrows from Q and P to the y-axis. (See Method 2.)

Notes:
- Point out that the y-coordinates of P and Q are the same, so they form a horizontal line that crosses the y-axis at $y = 3$.
- P and Q are in different quadrants. The absolute values of the x-coordinates tell us how far away each point is from the y-axis. The sum of these two distances gives us the distance between the points.

Have students do Try It! 10 on their own and discuss.

Notes:
- Ask students to look at the coordinates of the points first to determine whether the points are in the same or different quadrants, and whether the line is horizontal or vertical.
- Encourage struggling students to draw the coordinate graph and use Method 1. During the discussion, help them see how they could use Method 2 also.

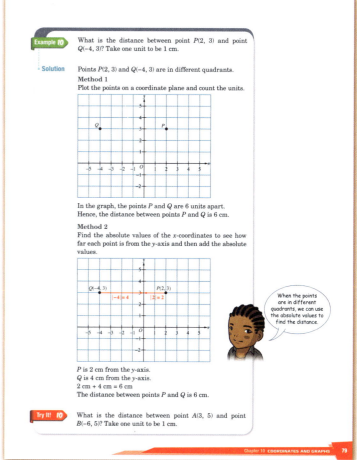

Student Textbook page 79

Try It! 10   Answers

Method 1

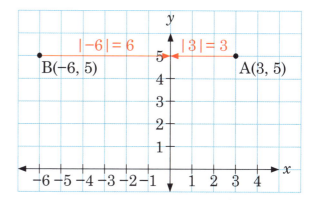

A and B are 9 units apart; 9 cm.

Method 2

A is 3 cm from the y-axis.
B is 6 cm from the y-axis.
$|3| + |-6| = 3 + 6 = 9$, 9 cm

2. Development

Have students study Example 11 and do Try It! 11.

Notes:
- In Try It! 11, encourage students to look at the coordinates of the points first to determine whether the points are in the same or different quadrants, and whether the line is horizontal or vertical.
- Encourage students to try to determine the distance without drawing the graph first by using Method 2 from Example 10. Then, have them draw the graph to check their answer.

Try It! 11   Answers

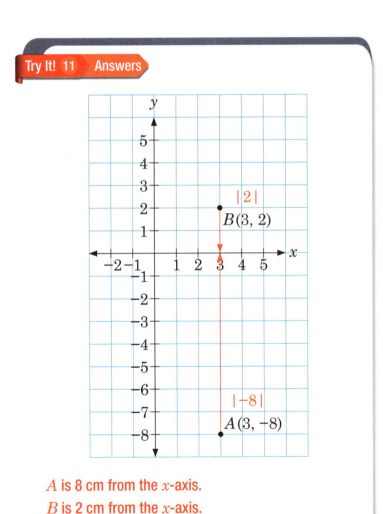

A is 8 cm from the x-axis.
B is 2 cm from the x-axis.
$|-8| + |2| = 8 + 2 = 10$, 10 cm

Student Textbook page 80

104

©2017 Singapore Math Inc. Dimensions Math® Teacher's Guide 6

3. Application

Have students study Example 12 and do Try It! 12.

Notes:

- The problems in Try It! 12 are similar to Example 9 in the previous lesson, except the coordinates of the vertices of the rectangle are in different quadrants, so we use addition to find the distance between the points.
- Encourage students to think about the lines the points will form before they draw the rectangle on a coordinate graph. In Example 12:
  - Points A and B, and points C and D, have the same y-coordinates, so they must form horizontal lines. The y-coordinates of A and B are positive and the y-coordinates of C and D are negative, so line segment CD must be below line segment AB.
  - Points A and D and points B and C have the same x-coordinates, so they must from vertical lines. The x-coordinates of A and D are negative and the x-coordinates of B and C are positive, so line segment AD must be to the left of line segment BC.

Try It! 12   Answers

See coordinate plane to the right.

Length $PQ = |-5| + |8| = 5 + 8 = 13$ cm
Width $PS = |4| + |-6| = 10$ cm
Area $= 13$ cm $\times 10$ cm $= 130$ cm^2
Perimeter $= (13 + 10) \times 2 = 46$ cm

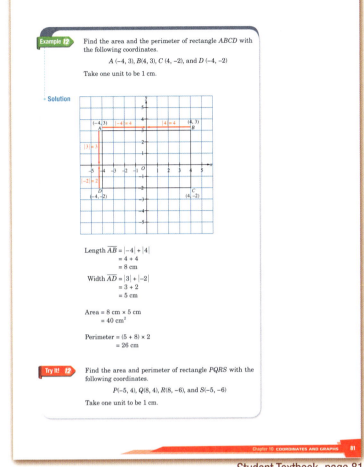

Student Textbook page 81

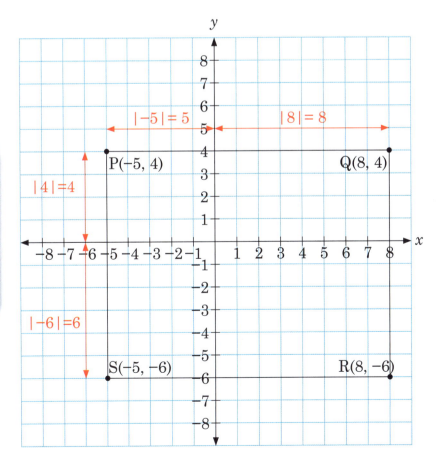

4. Extension

Have students cover the solution and do Example 13. Point out that two units of the graph paper equal one unit on the scale for both axes. Have students solve the problem and share and discuss their methods. Then, look at the solution given in the textbook and discuss.

Notes:
- The method shown in the textbook is based on finding the length of a side and then using that distance to complete side CD. This is based on the property that all sides of a square are the same length. Only the length of one side is needed for this method.
- Students may visualize where point D should go to form the square since side CD and side AD must be perpendicular.
- Encourage students to think about what the x- and y-coordinates of point D should be and why. In other words, point D should have the same x-coordinate as A and the same y-coordinate as C.

Have students do Try It! 13 on their own and discuss the solution.

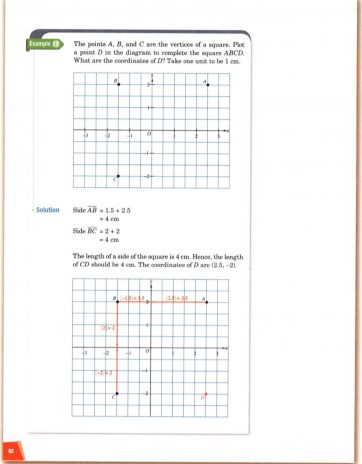

Student Textbook page 82

Notes:

- Remind students of the side properties of parallelograms. The lengths of the opposite sides are equal, and opposite sides are parallel.
- Since QP is a horizontal line segment that is 3 cm long, and QP and RS must be parallel, RS must also be a 3-cm horizontal line. A 3-cm line can be drawn from point R to determine the location of point S.

As an extension, ask students to find the area of the parallelogram.

- The height of the parallelogram $QPSR$ is perpendicular to the side RS, which is the distance between the side RS and QP and the segment along the y-axis from point S to the top of side of the parallelogram.
Height = $|2.5| + |-2|$ = 4.5 cm
Base (RS) = $|-3|$ = 3 cm
Area of parallelogram = base × height
$\qquad\qquad\qquad\qquad$ = 3 cm × 4.5 cm
$\qquad\qquad\qquad\qquad$ = 13.5 cm^2

Try It! 13 Answers

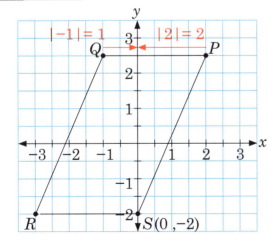

Side $QP = |-1| + |2| = 1 + 2 = 3$ cm
$RS = QP$ so $RS = 3$ cm
The value of the x-coordinate of R is -3, so the value of the x-coordinate of S can be found by $-3 + 3 = 0$. Thus, the coordinates of S are $(0, -2)$.

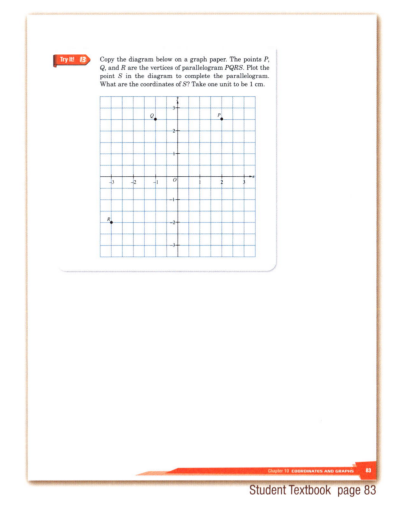

Student Textbook page 83

5. **Conclusion**

Summarize the important points of the lesson.

- When two points that form a horizontal or vertical line are in different quadrants, we can find the distance between the points by adding the absolute values of the coordinates that are different.
- We can apply this to find the lengths of the sides and areas of rectangles whose vertices are in different quadrants and whose sides are parallel to the axes.
- We can also apply it to determine the location of a missing vertex in other quadrilaterals such as squares and parallelograms, if we are given the coordinates of three of the vertices.

★ **Workbook: Page 44**

Lesson 6

Objective: Consolidate and extend the material covered thus far.

Have students work together with a partner or in groups. Students should try to solve the problems by themselves first, then compare solutions with their partner or group. If they are confused, they can discuss together.

Observe students carefully as they work on the problems. Give help as needed individually or in small groups.

⚙ BASIC PRACTICE

1. (a) Yes, because the y-coordinates are the same.
   $|10| - |2| = 10 - 2 = 8$ cm

   (b) Yes, because the y-coordinates are the same.
   $|-15| - |-4| = 15 - 4 = 11$ cm

   (c) Yes, because the y-coordinates are the same.
   $|12| - |3| = 12 - 3 = 9$ cm

   (d) No, because the y-coordinates are different.

2. (a) No, because the x-coordinates are different.

   (b) Yes, because the x-coordinates are the same.
   $|9.5| - |3.5| = 9.5 - 3.5 = 6$ cm

   (c) Yes, because the x-coordinates are the same.
   $|-10| - |-3| = 10 - 3 = 7$ cm

   (d) Yes, because the x-coordinates are the same.
   $|-12| - |-4| = 12 - 4 = 8$ cm

3. $PQ = |-15| + |8| = 15 + 8 = 23$ cm
   $RS = |-18| + |4.5| = 18 + 4.5 = 22.5$ cm
   $TU = |11| + |-9| = 11 + 9 = 20$ cm
   $VW = |10.6| + |-6.4| = 10.6 + 6.4 = 17$ cm

EXERCISE 10.2

In this exercise, take one unit in the coordinate plane to be 1 cm.

⚙ BASIC PRACTICE

1. For the pair of points shown in each diagram below, determine if the points are the endpoints of a horizontal line segment. If so, what is the distance between the pair of points?

   (a) $A(2, 9)$ $B(10, 9)$

   (b) $C(-15, 8)$ $D(-4, 8)$

   (c) $E(3, -7)$ $F(12, -7)$

   (d) $G(-6, -4)$ $H(-1, -5)$

2. For the pair of points shown in each diagram below, determine if a line segment drawn between them is vertical. If so, what is the distance between the pair of points?

   (a) $F(6, 11)$ $G(5, 2)$

   (b) $H(-4, 9.5)$ $G(-4, 3.5)$

   (c) $J(3, -3)$ $K(3, -10)$

   (d) $L(-6, -4)$ $M(-6, -12)$

Student Textbook page 84

⚙ FURTHER PRACTICE

4. (a) Horizontal line, No

   (b) Vertical line, No

   (c) Vertical line, Yes

   (d) Horizontal line, No

5. (a) $|-11| - |-2| = 11 - 2 = 9$ cm

   (b) $|-9| - |-1| = 9 - 1 = 8$ cm

   (c) $|3.5| - |2.5| = 3.5 - 2.5 = 1$ cm

6. (a) $|-12| - |-3| = 12 - 3 = 9$ cm

   (b) $|7| - |3| = 7 - 3 = 4$ cm

   (c) $|-12| - |-3| = 12 - 3 = 9$ cm

7. (a) |−8.4| + |4.8| = 8.4 + 4.8 = 13.2 cm

   (b) |−9| + |2.5| = 9 + 2.5 = 11.5 cm

   (c) |−13| + |3| = 13 + 3 = 16 cm

8. (a) |−20| + |8| = 20 + 8 = 28 cm

   (b) |16.5| + |−3.5| = 16.5 + 3.5 = 20 cm

   (c) |−11.9| + |7.6| = 11.9 + 7.6 = 19.5 cm

9. |4| + |−2| = 4 + 2 = 6 cm
   Perimeter = 4 × 6 = 24 cm
   Area = 6^2 = 36 cm²

10. Length AB = |−8| − |−2| = 8 − 2 = 6 cm
    Width BC = |5| + |−4| = 5 + 4 = 9 cm
    Area = 6 × 9 = 54 cm²
    Perimeter = (9 + 6) × 2 = 30 cm

11. Length PQ = |−6| + |7| = 6 + 7 = 13 cm
    Width PS = |4| + |−6| = 4 + 6 = 10 cm
    Area = 13 × 10 = 130 cm²
    Perimeter = (13 + 10) × 2 = 46 cm

12. Length of MN = |7| + |−3| = 7 + 3 = 10 cm

    Area of $ABNM$ = $\frac{2}{3}$ × 60 = 40 cm²

    $10w = 40$
    $w = 4$
    $AM = 4$ cm
    The coordinates of A are (−3,4).
    The coordinates of B are (7,4).

    $BC = \frac{60}{10} = 6$ cm
    $NC = 6 − 4 = 2$ cm
    $MD = NC = 2$ cm
    The coordinates of C are (7,−2).
    The coordinates of D are (−3,−2).

13. Perimeter = 35 cm

    $(l + w) \times 2 = 35$

    $(l + w) = 35 \div 2$

    $ = 17.5$ cm

    The sum of the length and width of the rectangle must be 17.5, so students can draw different rectangles such as $l = 10$ $w = 7.5$ or $l = 10.5$ $w = 7$, etc. For example:

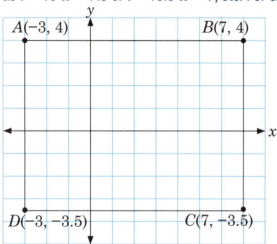

14. Length = $|-5| + |3| = 5 + 3 = 8$ cm

    $(8 + w) \times 2 = 28$

    $8 + w = 28 \div 2$

    $8 + w = 14$

    $ = 6$ cm

    The width has to be a distance of 6 on the y-axis from each point, so there are two possibilities: (–5, 2), (3, 2) or (–5, –10), (3, –10).

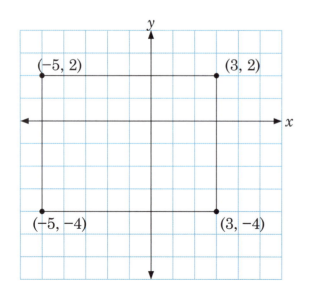

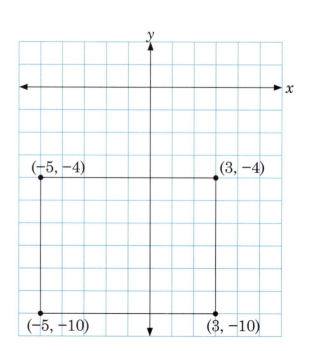

Lesson 7

Objective: Identify independent and dependent variables and describe their relationships.

1. Introduction

Read the examples on the bottom of page 86. Ask students to identify the quantities in each scenario:
- The distance Emily cycles and the number of calories she burns (e.g., if she cycles 5 miles, she burns 250 calories).
- The number of hours Bill works and the amount of pay he earns (e.g., if he works 10 hours, he makes $150).
- The amount of water used and the amount they have to pay for the water (e.g., if they use 500 cubic feet of water, they pay $20).

2. Development

Represent the scenarios on page 86 using two different variables.
- Emily cycles d miles and she burns c calories.
- Bill works h hours and earns m dollars.
- A household uses x cubic feet of water and pays $\$y$.

Have students read page 87 and discuss the independent and dependent variables for each scenario.

Notes:
- The independent variable can also be thought of as "the cause" and the dependent variable as "the effect." The effect is the result of the cause. For example:
Cause: Emily cycles. Effect: She burns calories.
Cause: Bill works. Effect: He earns money.
Cause: The household uses water. Effect: They have to pay for water used.
- The independent variable can be controlled. It does not depend on the other quantity. The dependent variable (the result) depends on the independent variable. For example, Emily can control how much she cycles. If she cycles a lot, she will burn a lot of calories. If she cycles a little, she will burn fewer calories.

Student Textbook page 87

3. Introduction

Have students do Example 14 (cover the solution) and work with partners or in groups to identify the independent and dependent variables in each scenario. Have students share and discuss their solutions, and the solutions given in the textbook.

Have students do Try It! 14 and share their solutions.

Try It! 14   Answers

(a) The number of brownies made depends on the number of guests who plan to attend the party, so g is the independent variable and b is the dependent variable.

(b) The gasoline bill depends on the number of miles driven, so x is the independent variable and y is the dependent variable.

(c) The amount of lemon juice depends on the number of lemons squeezed, so L is the independent variable and C is the dependent variable.

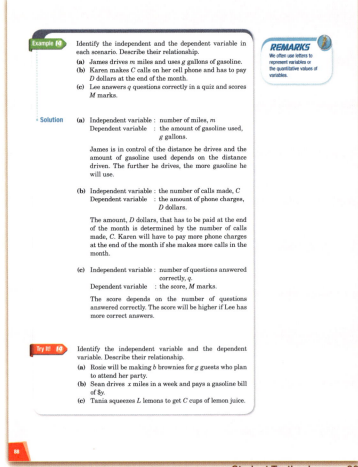

Student Textbook page 88

4. Extension

Have students work with partners or groups to think of scenarios where one quantity depends on the other quantity. Have them write statements for each situation (similar to Example 14) and identify the independent and dependent variables in each scenario.

Have students share their scenarios and discuss. Some possible examples:
- A person uses c cups of flour for a certain recipe and makes n cookies.
- A person eats c fewer calories daily and loses p pounds.
- A car drives m miles and uses g gallons of gasoline.
- A person saves $\$d$ each week and at the end of the month has $\$s$ in her savings account.

5. Conclusion

Summarize the main points of the lesson.
- In real-life situations, two quantities often change in relation to each other. We can represent these quantities with variables.
- The values of the independent variable are determined arbitrarily and can be controlled. The values of the dependent variable depend on the change in the independent variable.

★ Workbook: Page 57

Lesson 8

Objective: Represent relationships between variables using a table, and write an equation to represent this relationship.

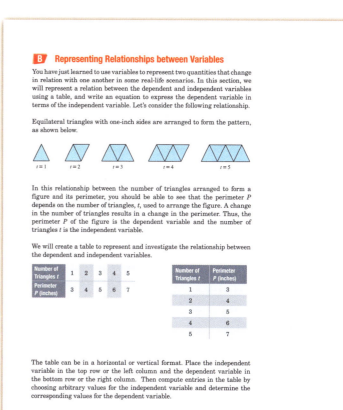

Student Textbook page 89

1. Introduction

In groups or pairs, give students pattern block triangles (equilateral triangles with 1-in sides) and have them make an equilateral triangle similar to the first one on page 89 ($t = 1$). (Alternatively, you can give students toothpicks or straws to form the triangles.)

Ask, "What is the perimeter of the triangle?" (3 in) Have students place another triangle on the right side of the first triangle so that there are two adjacent equilateral triangles like the second diagram on page 89 ($t = 2$).

Ask, "What is the perimeter of the shape?" (4 in)

Have students make 5 triangles and record the results in a horizontal table similar to the table on the left side of page 89.

No. of Triangles t	1	2	3	4	5
Perimeter P (in)	3	4	5	6	7

Ask, "What patterns can you see in the table?"
- The number of triangles increases by 1 each time.
- The perimeter increases by 1 each time we add a triangle.
- The perimeter is always 2 more than the number of triangles.

Ask, "What equation can you write to express this relationship between the number of triangles and the perimeter?" (In other words, "Express P in terms of t.")
- $P = t + 2$

Ask, "What is the independent variable and what is the dependent variable, and why?"
- The perimeter depends on the number of triangles, so t is the independent variable and P is the dependent variable.

2. Development

Read and discuss pages 89 – 90.

Note:
- Point out that the data can be recorded in a horizontal table or a vertical table.

Notes:
- Even though the quantities (perimeters) are changing every time we add a triangle, there is a fixed relationship between the number of triangles and the perimeters. Therefore, we can express the relationship with an equation.
- The values in the table can also be expressed as ordered pairs where t is the x-coordinate and P is the y-coordinate. (See top of page 90.) In the next lesson, we will plot these pairs to show the relationship between two quantities on a coordinate graph.

Ask, "What would be the perimeter for 100 triangles?"
- $P = t + 2$
  $P = 100 + 2$
  $P = 102$ in

3. Application

Have students study Examples 15 – 16 and do Try It! 15 – 16.

Notes:
- In Example 15, there is a vertical table. The independent variable is shown in the left column and the dependent variable in the right column. In Try It! 15, there is a horizontal table. The independent variable is shown in the top row and the dependent variable in the bottom row.
- Students can choose what kind of table they make but they should be familiar with both formats. (See REMARKS.)

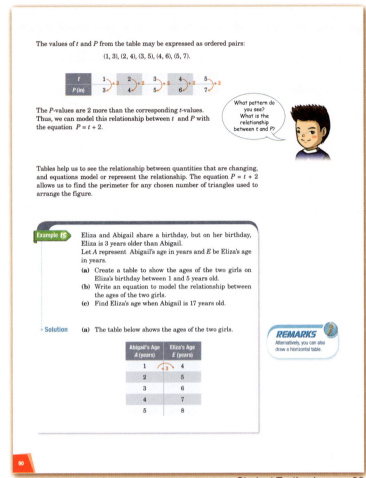

Student Textbook page 90

114

Try It! 15   Answers

(a)

Dog's age (*d* years)	3	4	5	6	7
Cat's age (*c* years)	1	2	3	4	5

(b) $c = d - 2$

(c) $c = 10 - 2 = 8$; 8 years old

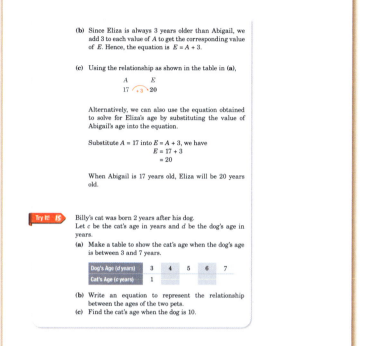

Student Textbook page 91

For Try It! 16, students can choose the kind of table they want to make, horizontal or vertical.

Try It! 16   Answers

(a) The total amount of money she earns depends on the number of hours she works, so h is the independent variable and p is the dependent variable.

(b) $p = 12.50h$ or $p = 12.5h$

(c)

Number of hours she babysits h	2	3	4	5	6
Amount of money by babysitting $\$p$	25	37.5	50	62.5	75

(d) $p = 12.50h$
$p = 12.50 \times 15$
$p = 187.50$
Her total earnings will be $187.50.

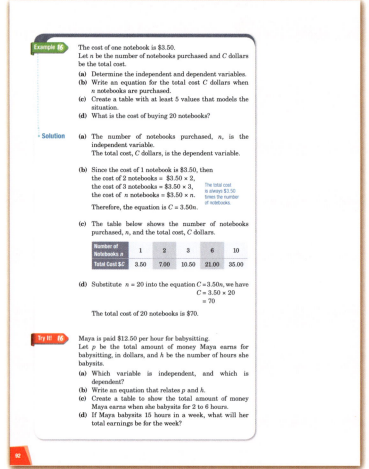

Student Textbook page 92

4. **Extension**

Have students study Example 17 and do Try It! 17.

Note:

- In these problems, the solution to (d) must be interpreted. In Example 17 (d), she will save exactly $85 in 11 weeks, but this is usually not the case in real life. If the shoes cost $86, then it would take her 12 weeks.

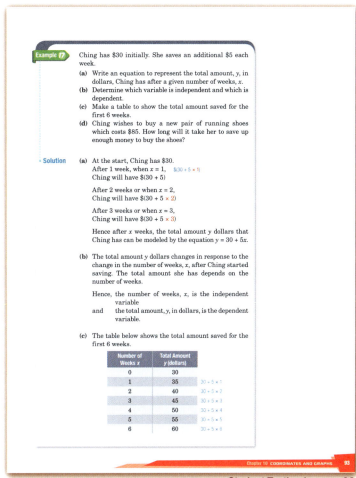

Student Textbook page 93

Notes:

- The solution for Example 17 (d) shows the equation written with y on the right side to make it easier to solve. However, it can be solved without moving y as:

    $y = 30 + 5x$
    $85 = 30 + 5x$
    $55 = 5x$
    $x = 11$

- Try It! 17 (d): She cannot buy 6.4 muffins. If she buys 7, it will cost more than $20. Thus, she can only buy 6 muffins.

Try It! 17 Answers

(a) The total cost depends on the number of muffins she buys, so n is the independent variable and c is the dependent variable.

(b) $c = 4 + 2.50n$ or $c = 4 + 2.5n$

(c)

Number of Muffins n	3	4	5	6	7	8
Total Cost of Purchase $\$c$	11.50	14	16.50	19	21.50	24

(d) **Method 1**

$4 + 2.5n = 20$
$2.5n = 16$
$n = 6.4$

She can buy 6 muffins.

Method 2

The table in (c) shows that it will cost more than $20 to buy 7 muffins, so she can buy 6 muffins.

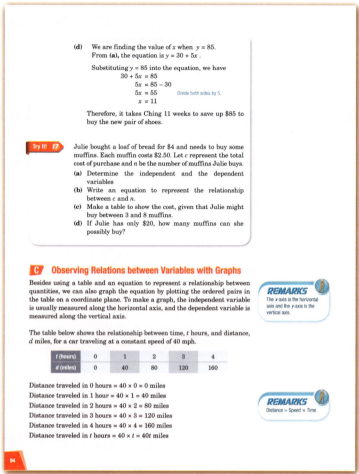

Student Textbook page 94

5. Conclusion

Summarize the main points of the lesson.

- We can represent quantities that change together in tables and with equations.
- An equation can help us solve problems involving finding one quantity given any value of the other quantity.

★ **Workbook: Page 58**

Lesson 9

Objective: Graph equations and solve problems involving changes in quantities.

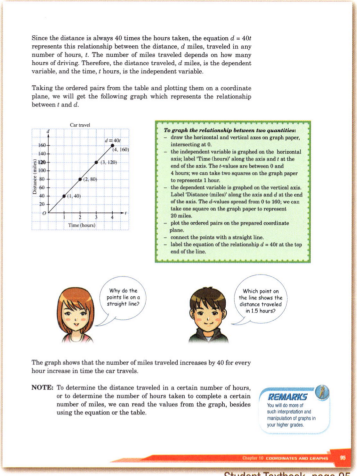
Student Textbook page 95

1. Introduction

Pose the following problem:

The table shows the relationship between time, t hours, and distance, d miles, for a car traveling at a constant speed of 40 mph.

t (hours)	0	1	2	3	4
d (miles)	0	40			

Ask students to find the distance (d) for 2, 3, and 4 hours and discuss how they determined the values for d.
- Distance = Speed × Time
- Distance = 40 × Number of Hours

t (hours)	0	1	2	3	4
d (miles)	0	40	80	120	160

Ask students to write an equation to show this relationship and discuss.
- $d = 40t$

Ask students how we could use the equation to find how far the car would travel in 12 hours.
- $d = 40 \times 12$
- $d = 480$ miles

2. Development

Ask students to draw a coordinate graph, with t as the x-axis and d as the y-axis (see graph on page 95), and have them plot the pairs of values from the table on the graph.
- The coordinate pairs are (0, 0), (1, 40), (2, 80), (3, 120), and (4, 160).

Ask students to look at the x and y values to determine the scales for the graph. Ask students what they notice about the graph.
- The points form a straight line.
- It shows the distance traveled for any amount of time (e.g., the distance for 1.5 hours would be 60 miles).

Read and discuss pages 94 – 96.

We know t is the x-coordinate and d is the y-coordinate because the distance traveled depends on the amount of time the car is driven. Thus, t is the independent variable and d is the dependent variable.

3. Application

Have students study Example 18 and do Try It! 18.

Notes:
- Students learned about unit rate in Chapter 6. In Example 18, the unit rate is given (30 words per minute).
- Students may notice that if you divide y by x you get the coefficient of x. This is true for direct proportions (e.g., Examples 18 and 19), but not for relationships that are not direct proportions (e.g., Example 20). Students will learn more about this in **Dimensions Math® 7**.
- We connect the points with a line because we can also determine the number of words typed for fractional amounts of minutes (e.g., 3.5 min, when all the points on the line are part of the solution).

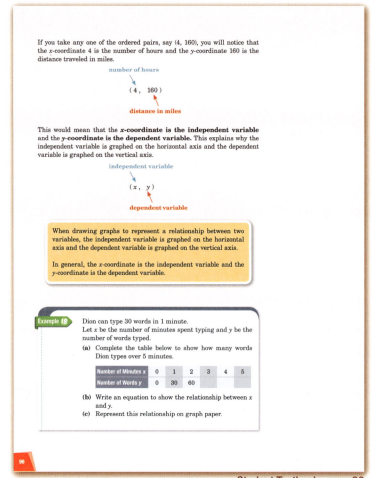

Student Textbook page 96

Notes:

- In Try It! 18, the water flows at a rate of 10 gallons every 5 minutes, so in 1 hour there will be 120 gallons of water in the pool.

  5 min $\rightarrow$ 10 gal

  1 min $\rightarrow \frac{10}{5} = 2$ gal

  60 min $\rightarrow 2 \times 60 = 120$ gal

  Students may know that there are 12 five-minute periods in an hour, so 12×10 gallons = 120 gallons.

Try It! 18   Answers

(a)

Time (t hours)	1	2	3	4	5
Volume V (gallons)	120	240	360	480	600

(b) $V = 120t$

(c) **Rate of Water Flowing into a Pool**

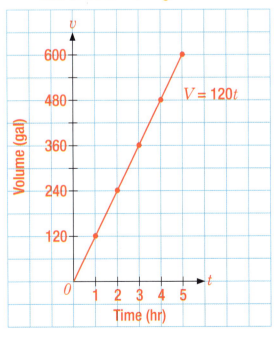

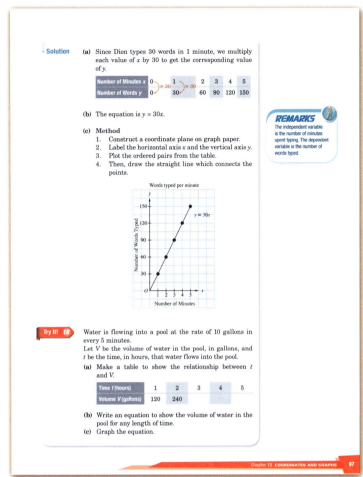

Student Textbook page 97

4. Extension

Have students study Example 19.

Discuss why the points are not connected with a line. (See note on Example 19 (d).)

Note:
- Even though we do not connect the points with a line here, if students want to use the graph to find perimeters beyond 6 steps, they could draw a line that extends farther up to the right. In that case, only the points with whole number values would be valid solutions.

Have students talk about the DISCUSS on the bottom of page 99. Then, have them do Try It! 19.

The dots on the graph should not be connected with a line, because you can only buy whole numbers of bags of buns.

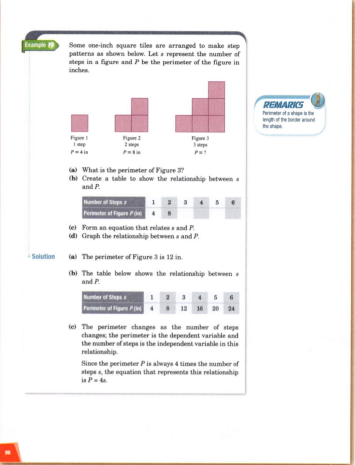

Student Textbook page 98

5. Conclusion

Summarize the main points of the lesson.
- We can show the relationship between two quantities by making a table and graphing the ordered pairs on a coordinate graph.
- We normally graph the independent variable on the x-axis and the dependent variable on the y-axis.
- If only discrete points (certain values) are included in the solution, we do not connect the points with a straight line.

Try It! 19 Answers

(a) The number of buns is always 6 times the number of bags.
$y = 6x$

(b)
Number of Bags x	1	2	3	4	5
Total Number of Buns y	6	12	18	24	30

(c) Total Number of Buns Sold

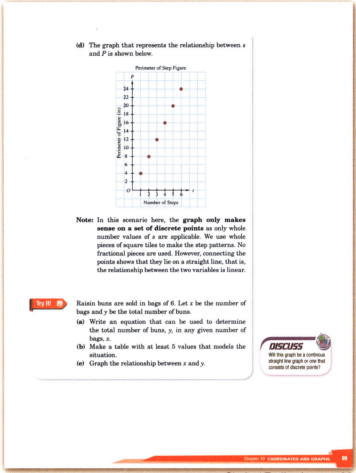

Student Textbook page 99

Notes:

- In Example 20, it may be helpful to begin the table with 0 weeks and write an expression for each week to understand the pattern better.

No. of Weeks	Expression	Total No. of Stamps
0	2 + 3 × 0	2
1	2 + 3 × 1	5
2	2 + 3 × 2	8
3	2 + 3 × 3	11
4	2 + 3 × 4	14
5	2 + 3 × 5	17
6	2 + 3 × 6	20
7	2 + 3 × 7	23

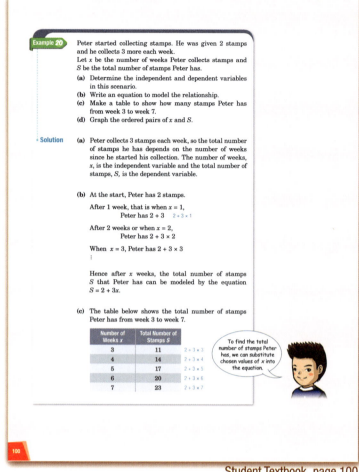

Student Textbook page 100

Note:
- For the graph in Example 20 (d), points could also be plotted for weeks 0, 1, and 2, (0, 2), (1, 6), and (2, 8).
- When drawing graphs in Try! It 20:
  - Example 15 is on page 90.
  - Try It! 17 is on page 94. See table from Try It! 17 (c) on page 250 for the coordinate pairs.

Try It! 20 Answers

(a) **Abigail's and Eliza's Ages on their Birthday**

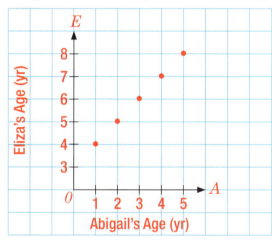

(b) **Cost of Bread and Muffins**

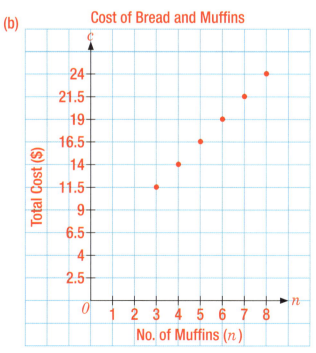

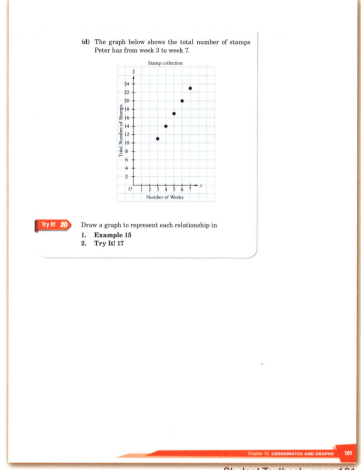

Student Textbook page 101

★ **Workbook: Page 61**

Lesson 10

Objective: Consolidate and extend the material covered thus far.

Have students work together with a partner or in groups. Students should try to solve the problems by themselves first, then compare solutions with their partner or group. If they are confused, they can discuss together.

Observe students carefully as they work on the problems. Give help as needed individually or in small groups.

BASIC PRACTICE

1. (a) Independent variable: number of years
   Dependent variable: growth in height of boy

   (b) Independent variable: number of burgers
   Dependent variable: total cost of burgers

   (c) Independent variable: number of correct answers
   Dependent variable: test score

   (d) Independent variable: number of donuts
   Dependent variable: number of calories gained

2. (a) The amount he gets paid depends on the number of bottles he sells, so x is the independent variable and P is the dependent variable.

   (b) The number of cookies she can make depends on the amount of flour she has, so F is the independent variable and k is the dependent variable.

   (c) The amount of money he should bring depends on the number of days they will go on vacation, so D is the independent variable and M is the dependent variable.

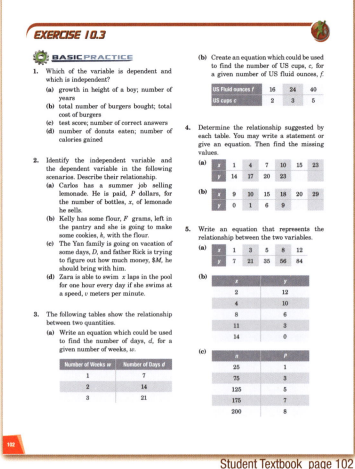

Student Textbook page 102

(d) The number of laps she swims depends on her speed, so v is the independent variable and x is the dependent variable.
Note: Students may think that x is the independent variable because when they created graphs the x-axis was always for the independent variable. Help them see that since this is just a convention, they need to look at the relationship to determine which variable is independent or dependent.

3. (a) The number of days is always 7 times the number of weeks. $d = 7w$

   (b) The number of cups is equal to the number of fluid ounces divided by 8. $c = \frac{f}{8}$ or $c = \frac{1}{8}f$

4. (a) $y = x + 13$

   y is always 13 more than x.

x	1	4	7	10	15	23
y	14	17	20	23	28	36

   (b) $y = x - 9$

   y is always 9 less than x.

x	9	10	15	18	20	29
y	0	1	6	9	11	20

5. (a) To find y, multiply x by 7. $y = 7x$

   (b) To find y, subtract x from 14. $y = 14 - x$

   (c) To find P, divide n by 25. $P = \frac{n}{25}f$ or $P = \frac{1}{25}n$

6. (a) Independent variable: w
   Dependent variable: A

w	A
0	0
1	4
3	12
4	16
6	24

   (b) Independent variable: x
   Dependent variable: y

x	0	1	2	5	7
y	5	8	11	20	26

7. (a)

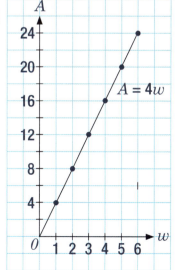

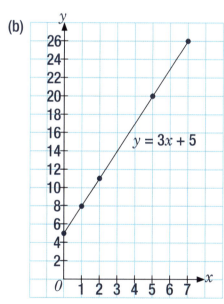

FURTHER PRACTICE

8. (a)

x	y
1	50
2	100
3	150
4	200
6	300

$y = 50x$

(b)

x	0	2	4	8	10
y	2	4	6	10	12

$y = x + 2$

9. (a)

Sonia's Age S (years)	4	5	6	7	8
John's Age J (years)	0	1	2	3	4

(b) $J = S - 4$ or $S = J + 4$

(c) $J = 16 - 4 = 12$; 12 years old

(d) **Relationship between Ages**

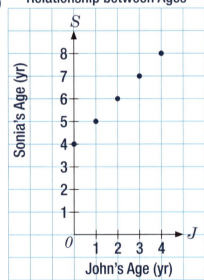

10. (a) Independent variable: number of bags b
    Dependent variable: total cost in dollars, C

(b) $C = 3.5b$

(c) $59.50

Student Textbook page 104

(d)

Bags of Apples, b	1	2	3	4	5
Total Cost C dollars	3.50	7.00	10.50	14.00	17.50

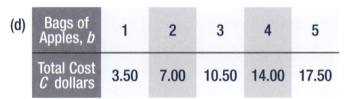

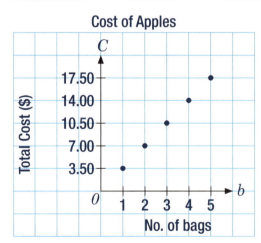

There is no line connecting the points because you can only buy a whole number of bags of apples.

11. (a)

Number of Hours x	1	2	3	4	5
Amount Paid $\$y$	20	40	60	80	100

(b) $y = 20x$

(c)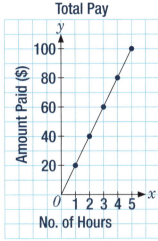

There is a line connecting the points because he could possibly work a fraction of an hour, such as 4.5 hours. If he can only work a whole number of hours, however, there would be no line.

12. (a) $C = 8 + r$

(b) Independent variable: number of rides r;
Dependent variable: total cost in dollars, C

(c)

Number of Rides r	2	3	4	5	6	7
Total Cost $\$y$	10	11	12	13	14	15

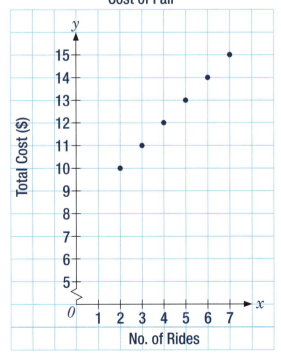

MATH @ WORK

13. $y = 15x$

Independent variable: number of gallons of gasoline x
Dependent variable: number of miles the truck drives, y

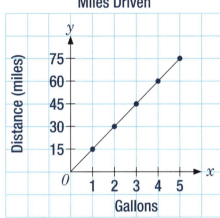

14. Independent variable: number of miles driven, x
    Dependent variable: total cost in dollars, C
    $C = 6 + 1.5x$; 11 miles

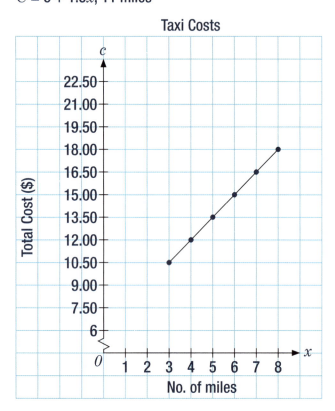
Taxi Costs

15. $y = \$0.50x + \$10.00$

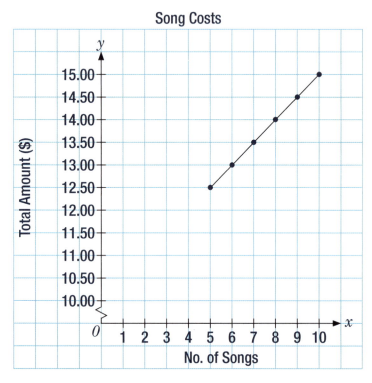
Song Costs

BRAIN WORKS

16. (a)

Number of Minutes, x	10	20	30	40	50
Number of Cupcakes, y	4	8	12	16	20

Multiples of 10 are easy for the x values but students can choose different numbers of minutes for their table.

(b) If you divide the value of x by the value of y, you get $\frac{2}{5}$.

$y = \frac{2}{5}x$

(c) No. of Cupcakes Luz Can Decorate

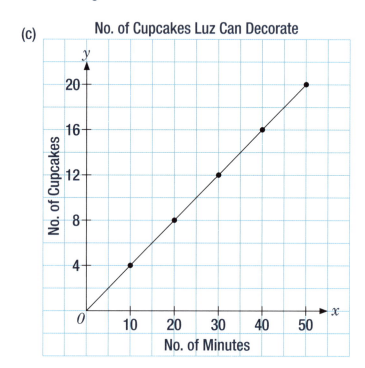

(d) We can use the equation in (b) to determine the number of cupcakes Luz can decorate.

(i) $\frac{2}{5} \times 35 = 14$

(ii) $\frac{2}{5} \times 42.5 = 0.4 \times 42.5 = 17$

(iii) $\frac{2}{5} \times 110 = 44$

17.   1 box ⟶ $4

2 boxes ⟶ 2 × $4 = $8

You would only get 2 boxes if you decide you do not want the free box.

3 boxes ⟶ 2 × $4 = $8

When you buy 2 boxes, you get 1 box free.

4 boxes ⟶ 2 × $4 + $4 = $12

When you buy 2 boxes, you get 1 free, so you will have 3 boxes. Then you only have to pay $4 for another box.

5 boxes ⟶ 4 × $4 = $16

If you buy 4 boxes, you get 2 free boxes, so you would have 6 boxes. You would only get 5 boxes if you decide you do not want the sixth free box.

6 boxes ⟶ 4 × $4 = $16

If you buy 4 boxes, you get 2 free boxes.

7 boxes ⟶ 4 × $4 + $4 = $20

If you buy 4 boxes, you get 2 free boxes. Then you need to pay $4 for 1 more box.

Number of Boxes Bought b	Amount Paid P dollars
1	4
2	8
3	8
4	12
5	16
6	16
7	20

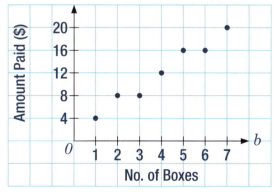

It looks this way because different x-values can have the same y-value. It is not a linear relationship.

18. (a) Every point on the x-axis has a y-coordinate of 0.

x	0	1	2	3
y	0	0	0	0

$y = x \times 0$, thus the equation for the x-axis is $y = 0$.

(b) Every point on the y-axis has an x-coordinate of 0.

x	0	0	0	0
y	1	2	3	4

$x = y \times 0$, thus the equation for the y-axis is $x = 0$.

Lesson 11

Objective: Summarize and reflect on important ideas learned in this chapter, and solve a non-routine problem.

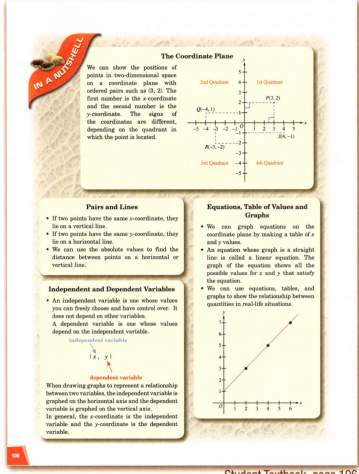

Student Textbook page 106

Note:
- This lesson could be done in class or assigned for students to do independently at home or at school.

1. In a Nutshell

Use this page to summarize the important ideas learned in this chapter.

Give examples where needed.

2. Write in Your Journal

Have students do the writing activity and share their answers. Answers will vary.

Note:
- In the 1st quadrant, the coordinates will both be positive. In the 3rd quadrant, the coordinates will both be negative. Thus, points with the same x and y coordinates, such as (3, 3) and (−5, −5), would have to be in one of these two quadrants.

3. Extend Your Learning Curve

This activity can be completed in class or done as an independent assignment.

Note:
- To get from P to Q, you go to the right 6 units and up 12 units. To find $\frac{1}{3}$ of the distance:

$6 \times \frac{1}{3} = 2, \quad 12 \times \frac{1}{3} = 4$

or

$\frac{6 \div 3}{12 \div 3} = \frac{2}{4}$

Thus, to go $\frac{1}{3}$ of the distance, you would go right 2 units and up 4 units, so the coordinates of R are (5, 7).

To find the coordinates of S, start from point Q and go 2 units left and 4 units down, so S is (7, 11).

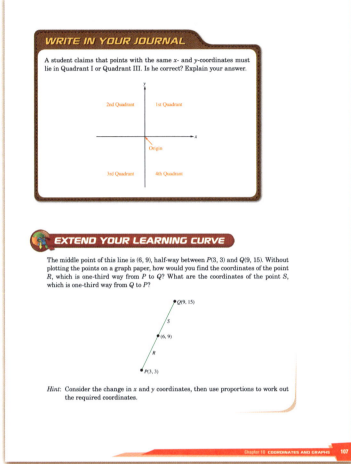

Student Textbook page 107

Notes

Chapter 11: Area of Plane Figures

Lesson	Objectives	Class Periods	Textbook & Workbook	Teacher's Guide Page	Additional Materials Needed
1	• Determine the base and height of a parallelogram. • Derive the formula for the area of a parallelogram.	1	TB: 108–114	140	1-cm grid paper, rulers, scissors, index cards, blank copy paper, dynamic geometry software (optional)
2	• Find the area of parallelograms. • Find the base or height of a parallelogram given the area and the base or height.	1	TB: 114–118 WB: 72–77	146	1-cm grid paper, rulers, scissors, blank copy paper, index cards
3	• Consolidate and extend the material covered thus far.	1	TB: 119–122	151	Tangram template (p. 156)
4	• Determine the base and height of a triangle. • Derive the formula for the area of a triangle.	1	TB: 123–126	157	Blank paper, 1-cm grid paper, index cards, rulers, scissors
5	• Find the area of triangles. • Find the base or height of a triangle given the area and the base or height.	1	TB: 126–130 WB: 78–83	161	Blank paper, index cards, rulers, scissors
6	• Find the area of composite plane figures that are composed of parallelograms and triangles.	1	TB: 130–134 WB: 84–87	167	1-cm grid paper, rulers
7	• Consolidate and extend the material covered thus far.	1	TB: 135–138	171	
8	• Determine the base and height of a trapezoid. • Derive the formula for the area of a trapezoid and find the area of trapezoids.	1	TB: 139–146 WB: 88–92	177	1-cm grid paper, rulers

Continues on next page.

Chapter 11: Area of Plane Figures

Lesson	Objectives	Class Periods	Textbook & Workbook	Teacher's Guide Page	Additional Materials Needed
	• Find one base or height of a parallelogram given the area, the length of one base and the height, or find the height given the area and the sum of the bases.				
9	• Consolidate and extend the material covered thus far.	1	TB: 147–150	185	
10	• Summarize and reflect on important ideas learned in this chapter, and solve a non-routine problem.	1	TB: 151–152	189	

In elementary school, students learned about square units and how to find the areas of rectangles and squares by utilizing multiplication to establish formulas. They also learned to find the areas of composite figures formed of rectangles in different ways. For example:

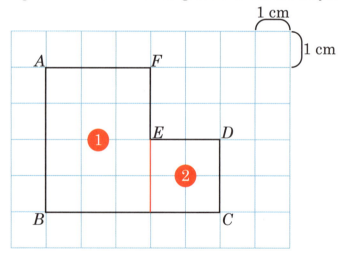

Method 1:

Cut figure ABCDEF into two rectangles.

Area of rectangle 1 = 3 cm × 4 cm = 12 cm²
Area of rectangle 2 = 2 cm × 2 cm = 4 cm²
Area of figure ABCDEF = 12 cm² + 4 cm² = 16 cm²

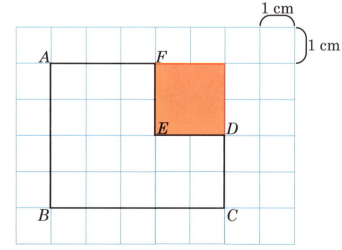

Method 2:

Add a small square to create a larger rectangle.

Area of large rectangle = 5 cm × 4 cm = 20 cm²
Area of square = 2 cm × 2 cm = 4 cm²
Area of figure ABCDEF = 20 cm² − 4 cm² = 16 cm²

Chapter 11: Area of Plane Figures

- In Method 1, we notice that the area of the composite figure is the sum of the areas of the two distinct rectangles.
- In Method 2, we notice that the area of the figure can be found by finding the area of a larger rectangle and subtracting the area not included in the original figure.

In this chapter, students will extend these ideas to find the areas of parallelograms, triangles, trapezoids, and compound figures that are composed of these shapes.

By examining how their methods are related to the formulas for the areas of previously learned figures, students can develop formulas for finding the area of each new shape. For example, to find the area of a parallelogram, we notice that we can rearrange portions of the figure. By cutting off a right triangle from one side and moving it to the other side, we can form a rectangle with the same area as the parallelogram.

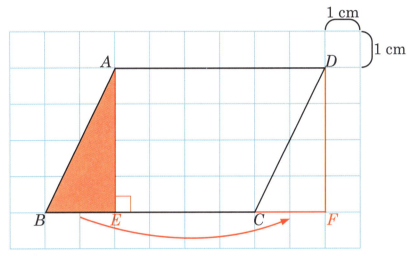

The height of parallelogram $ABCD$ is the same as the height of rectangle $AEFD$. The bases of the parallelogram and the rectangle have the same length. Thus, to find the area of the parallelogram, we can multiply the base by the height, where the height is perpendicular to the base.

Area of parallelogram = 6 cm × 4 cm = 24 cm²
Area of parallelogram = base × height

To find the area of a triangle, we can add another congruent triangle to create a parallelogram that has twice the area of the triangle. In this case, the parallelogram and the triangle do not have the same area. We need to multiply the area of the parallelogram by $\frac{1}{2}$ (or divide it by 2) to find the area of the triangle.

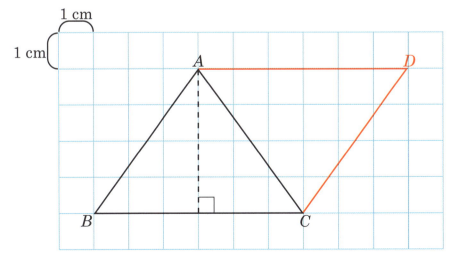

The base and height of triangle ABC are the same as the base and height of parallelogram $ABCD$.

Area of parallelogram = 6 cm × 4 cm = 24 cm²

Area of triangle = $\frac{1}{2}$ × (6 cm × 4 cm) = 12 cm²

Area of triangle = $\frac{1}{2}$ × Area of the parallelogram

Area of triangle = $\frac{1}{2}$ × base × height

Chapter 11: Area of Plane Figures

To find the area of a trapezoid, we can add a congruent trapezoid to create a parallelogram that has twice the area as the trapezoid.

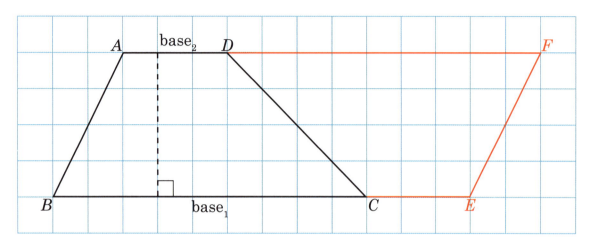

The base of the parallelogram is the sum of the top and bottom bases of the trapezoid.

Area of parallelogram = (9 cm + 3 cm) × 4 cm = 48 cm²

Area of trapezoid = $\frac{1}{2}$ × (9 cm + 3 cm) × 4 cm = 24 cm²

Area of trapezoid = $\frac{1}{2}$ × (sum of the bases) × height

Area of trapezoid = $\frac{1}{2}$ × (base$_1$ + base$_2$) × height

It is important to engage students in hands-on activities like cutting paper and moving parts to transform new shapes into shapes that students already know how to find the area of.

This 1-cm grid paper can be copied and distributed to students when needed for lessons or practice.

Lesson 1

Objectives:
- Determine the base and height of a parallelogram.
- Derive the formula for the area of a parallelogram.

1. **Introduction**

Discuss page 108.

State the learning goals of the chapter, LET'S LEARN TO …

Student Textbook page 108

2. **Development**

Give students a blank sheet of paper and ask them to fold it horizontally in half and then horizontally in half again.

Have them draw the heights in with a ruler and label the right angles similar to the second picture on page 109.

Read and discuss pages 109 – 110 and discuss the REMARKS.

Ask students to place the corner of an index card or the corner of a paper where the base and height meet to help them see that they are perpendicular (i.e., meet at right angles).

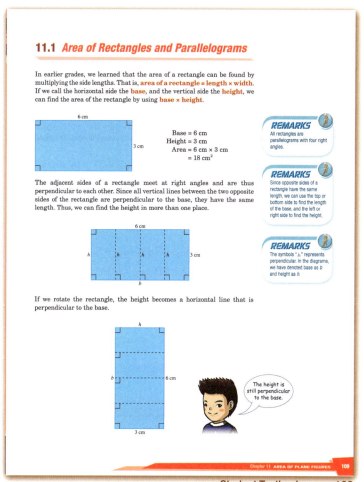

Student Textbook page 109

Notes:

- Since a parallelogram is a quadrilateral with two pairs of parallel sides, all rectangles are parallelograms. However, all parallelograms are not rectangles because the adjacent sides of rectangles are perpendicular to each other.
- Discuss how we use arrows to denote parallel sides. The two sides with one arrow are parallel and the two sides with two arrows are parallel.
- The base is simply the side that is perpendicular to a given height, and has nothing to do with the idea that an object rests on its base or that the base has to be horizontal to the bottom of the page.
- The height of a parallelogram is the perpendicular distance from one base to its opposite parallel side. The height can be drawn within or outside of the parallelogram. Help students see this by drawing the height in different places.

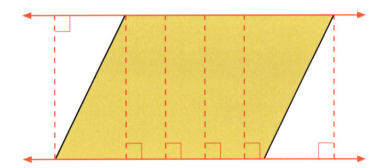

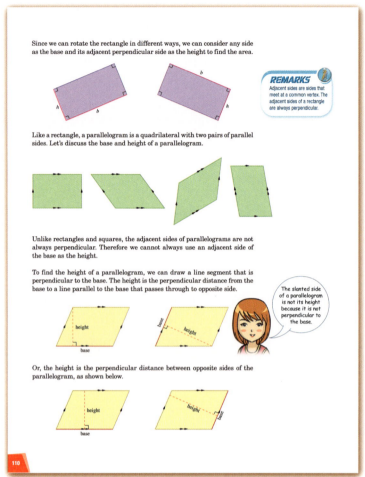

Student Textbook page 110

Have students study Example 1 and do Try It! 1.

Discuss the solutions and the REMARKS.

Notes:
- We usually name polygons in alphabetical order based on the vertices (e.g., parallelogram *ABCD*). It is not incorrect, however, to name it in another order as long as we name it in the order in which the points go around the shape. (In Example 1, we can say parallelogram *ADCB* but not *ACDB*.)
- It may be difficult for some students to see why *GH* in Example 1 is a height of the parallelogram. Ask students to turn the book so they can see base *AB* on the bottom and place the corner of an index card where the base and the height meet to see that they are perpendicular. They can also do this for Try It! 1 (d).

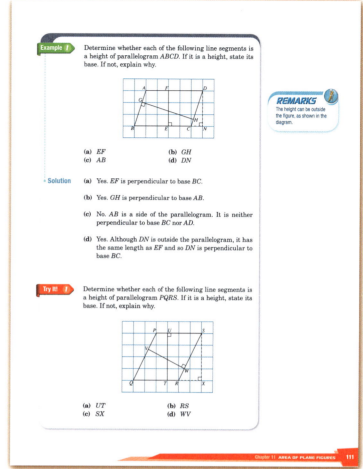

Student Textbook page 111

Try It! 1 Answers

(a) Yes. UT is perpendicular to base QR.

(b) No. RS is a side of the parallelogram. It is neither perpendicular to base QR nor PS.

(c) Yes. Although SX is outside the parallelogram, it has the same length as UT and is perpendicular to base QR.

(d) Yes. WV is perpendicular to base SR.

3. **Application**

Give students 1-cm grid paper and rulers, and have them complete problems 1 – 3 of Class Activity 1.

With a partner or in groups, have them answer and discuss problem 4.

Note: Since the bases and areas of the rectangle and parallelogram are the same length, the height must also be the same. Thus, the height of the parallelogram must be perpendicular to the base.

Answers for Class Activity 1

3. (a) Rectangle
   (b) $EF = 6$ cm, $AE = 4$ cm
   (c) 4 cm × 6 cm = 24 cm^2
4. (a) They are the same length.
   (b) The area remains the same because we just moved some of the area. We did not add or subtract area.
   (c) They have the same area.
   (d) Area of parallelogram = base × height

Notes:
- Help students see that for rectangles we can multiply the adjacent sides to find the area, because the side of a rectangle is perpendicular to the base. For non-rectangular parallelograms, the side is not perpendicular to the base, so we need to multiply by the perpendicular height.
- Remind students that cm^2 is a unit of area and it is read "square centimeters."

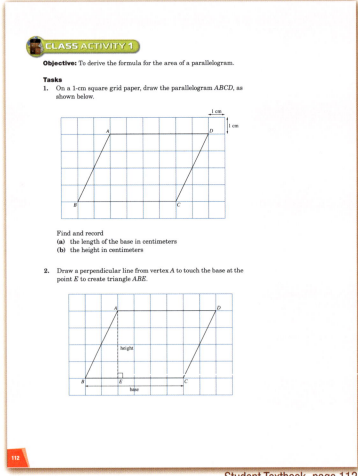

Student Textbook page 112

Ask students to share their conclusions from Class Activity 1.

Use the bottom of page 113 to summarize the conclusions from Class Activity 1.

4. **Extension**

Ask students to draw parallelogram *ABCD* for Class Activity 1 again on grid paper.

Tell students to draw a different parallelogram using the same base *BC* that has the same height as *ABCD*. (See below for examples.)

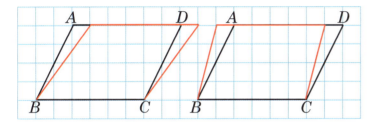

Ask students to share the different parallelograms they drew.

Ask, "Are the areas of the parallelograms the same or different? Why or why not?"
- The areas are the same because the base is the same (6 cm), and they have the same height (4 cm).
- They share the same bottom base. If we move the top base anywhere between the parallel lines, the height remains the same, thus, the areas remain the same.

Note: This last point can be shown using dynamic geometry software such Geometer Sketchpad or Cabri.

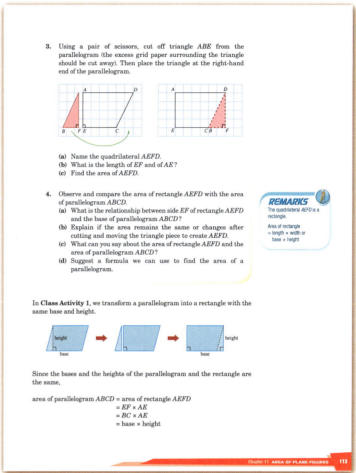

Student Textbook page 113

5. **Conclusion**

Summarize the important points of the lesson.
- A parallelogram is a quadrilateral with two pairs of parallel sides. A rectangle is also a parallelogram.
- The height of a non-rectangular parallelogram is not the side, but the perpendicular distance between the bases.
- We can find the area of a parallelogram by relating it to a rectangle that has the same base and height. To find the area of a parallelogram, multiply the base by the height.
- If two parallelograms have the same base and height, the areas are the same, even if they look different.

Lesson 2

Objectives:
- Find the area of parallelograms.
- Find the base or height of a parallelogram given the area and the base or height.

1. Introduction

Give students grid paper and ask them to draw parallelogram *ABCD* (shown below) with a base of 4 cm and height of 6 cm.

Note: Vertex *D* should be 6 cm horizontally to the right and 6 cm up from vertex *C*.

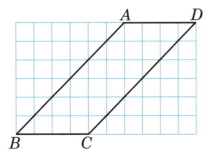

Ask, "Can we always construct a rectangle from any parallelogram? How about this parallelogram?"

Note: This is the same as the problem on page 114.

Ask students to try to change the parallelogram into a rectangle and then find the area. They can cut and paste or just shade and draw arrows showing how they would change the shape.

Notes:
- Since the height cannot be drawn within the shape from the base to its opposite side, they cannot just cut a triangle off of one end and move it to make a rectangle.
- If students cannot think of any methods, have them look at page 115.

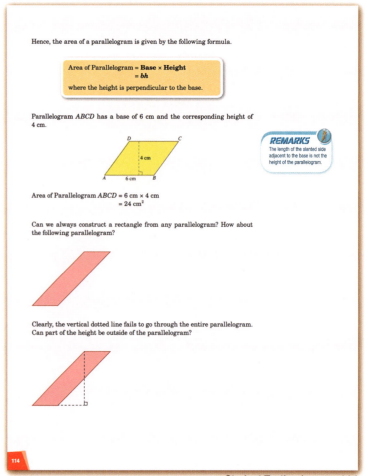

Student Textbook page 114

Ask students to share their methods. One possible method is to cut the parallelogram into two (or more) smaller parallelograms and cut and move a triangular portion to make rectangles.

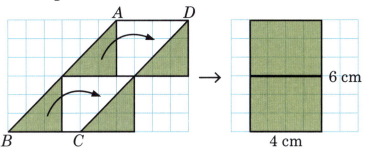

$4 \times 3 + 4 \times 3 = 4 \times (3 + 3) = 4 \times 6$

Area = 24 cm²

Notes:

- This method can be connected with the formula by writing out the equation:

$$(b \times \tfrac{1}{2} \times h) + (b \times \tfrac{1}{2} \times h) = 2 \times (b \times \tfrac{1}{2} \times h)$$
$$= 2 \times \tfrac{1}{2} \times b \times h$$
$$= b \times h$$

- The height of the parallelogram is 6 cm, which can be drawn outside the base by drawing a line from vertex D to an extension of the line representing the base. Alternatively, we can extend a horizontal line from vertex C to the right and draw the height by drawing a perpendicular line to vertex D.

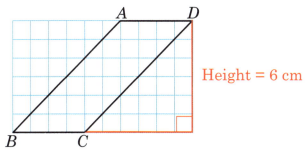

Height = 6 cm

Ask students to multiply the base by the height of the original parallelogram to see that the area is the same as the area of the rectangle.

base × height = 4 cm × 6 cm = 24 cm²

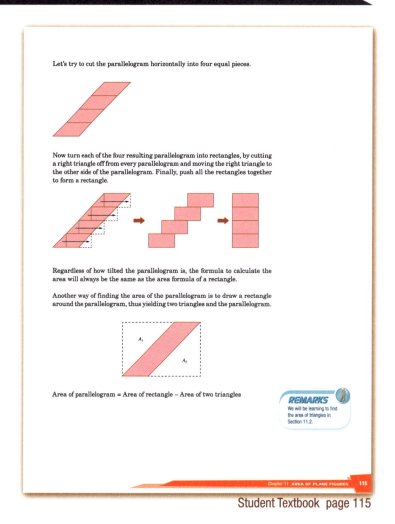

Student Textbook page 115

2. Development

Read and discuss pages 114 – 115. Compare and contrast the method given on page 115 with the methods the students came up with.

Point out that the sum of the heights of the rectangles is equal to the height of the parallelogram. Thus, we can still use base × height to find the area, even when the height is shown outside the base.

3. Application

Have students study Examples 2 – 3, do Try It! 2 – 3, and then discuss the solutions.

Example 2:
- Discuss the REMARKS on page 116. Remind students that any side of the parallelogram can be considered as the base.
- The dimensions of a side are also given. Discuss with students why the length of the side cannot be used as the height. (It is not perpendicular to the base.)

Try It! 2 Answers

(a) The base is NO and a corresponding height is RQ.
Area of parallelogram $MNOP$ =
7.5 cm × 2 cm = 15 cm²

(b) The base is DE and a corresponding height is HI.
Area of parallelogram $DEFG$ =
3 cm × 8.2 cm = 24.6 cm²

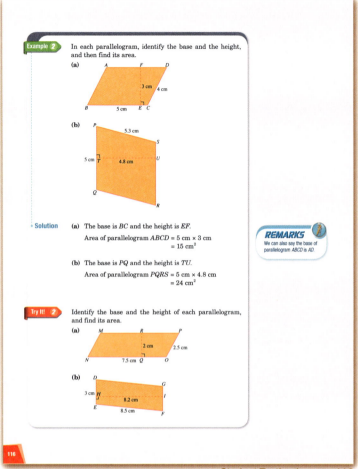

Student Textbook page 116

148

Example 3
- If students are confused about where the height is drawn in Example 3 (b), have them rotate the book so the base is on the bottom and use the corner of an index card to see that the height is perpendicular to the base.
- In Try It! 3 (a), the height extends out to the left. Remind students that the base of the parallelogram and its opposite side are parallel, and that the height can be found anywhere in between the bases, even outside of the base.

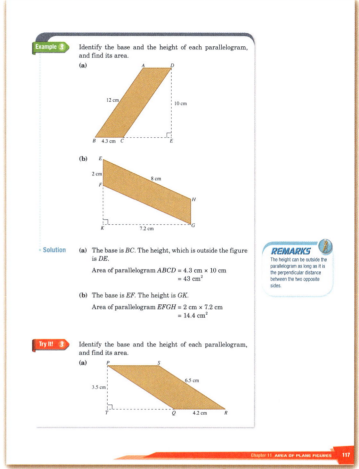

Student Textbook page 117

Try It! 3 Answers

(a) The base is QR and the height is PT.
Area of parallelogram $PQRS$ =
4.2 cm × 3.5 cm = 14.7 cm²

(b) The base is CD and the height is BE.
Area of parallelogram $ABCD$ =
8 cm × 19 cm = 152 cm²

4. Extension

Have students study Example 4, do Try It! 4, and then discuss the solutions.

Example 4:
- Here we are finding a missing dimension of the parallelogram given the area and one dimension. This is similar to what students did in earlier grades when they found the length of one side of a rectangle given its area and length of one side.
- The following relationships hold true for parallelograms:

    Area = Base × Height
    Base = Area ÷ Height
    Height = Area ÷ Base

Try It! 4   Answer

Area of cardboard = length of side × perpendicular distance

3,600 cm² = 80 cm × perpendicular distance

Perpendicular distance = $\frac{3{,}600 \text{ cm}^2}{80 \text{ cm}}$ = 45 cm

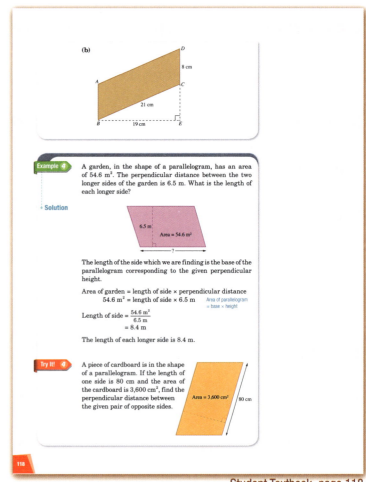

Student Textbook page 118

★ **Workbook: Page 72**

5. Conclusion

Summarize the main points of the lesson.
- We can use the formula base × height to find the area of any parallelogram, even when the height is drawn outside the base.
- Any side of the parallelogram can be considered the base. The height is always the perpendicular distance from the base to the opposite base (opposite parallel side).
- If we know the area of the parallelogram and one dimension (base or height), we can divide the area by the dimension we know to find the missing dimension.

Lesson 3

Objective: Consolidate and extend the material covered thus far.

Have students work together with a partner or in groups. Students should try to solve the problems by themselves first, then compare solutions with their partner or group. If they are confused, they can discuss together.

Observe students carefully as they work on the problems. Give help as needed individually or in small groups.

Note: Due to the complexity of the problems, students may use a calculator at the teacher's discretion, especially for problems 7 – 16.

 BASIC PRACTICE

1. (a) *DE* (b) *CH*
   (c) *CJ* (d) *DN*

2. (a) Base *AD*, height *BF*
   (b) Base *AD*, height *BG*
   (c) Base *BC*, height *DK*
   (d) Base *AD*, height *BM*

3. (a) 8 in × 5 in = 40 in²
   (b) 9 in × 11 in = 99 in²
   (c) 7 cm × 14 cm = 98 cm²
   (d) 12 cm × 17 cm = 204 cm²

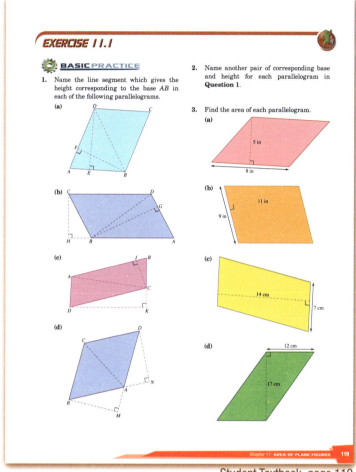

Student Textbook page 119

Chapter 11 AREA OF PLANE FIGURES

FURTHER PRACTICE

4. (a) $9 \text{ m} \times 6.4 \text{ m} = 57.6 \text{ m}^2$

   (b) $5.6 \text{ m} \times 8 \text{ m} = 44.8 \text{ m}^2$

   (c) $7 \text{ ft} \times 3\frac{1}{4} \text{ ft} = 22\frac{3}{4} \text{ ft}^2$

   (d) $2\frac{1}{2} \text{ ft} \times 4\frac{3}{4} \text{ ft} = 11\frac{7}{8} \text{ ft}^2$

   (e) $23 \text{ mm} \times 50 \text{ mm} = 1{,}150 \text{ mm}^2$

   (f) $32 \text{ mm} \times 34 \text{ mm} = 1{,}088 \text{ mm}^2$

Note: For 4 (e) and (f), students could also convert mm to cm to find the area:

   (e) $2.3 \text{ cm} \times 5 \text{ cm} = 11.5 \text{ cm}^2$

   (f) $3.2 \text{ cm} \times 3.4 \text{ cm} = 10.88 \text{ cm}^2$

5. $PN = \dfrac{112 \text{ cm}^2}{14 \text{ cm}} = 8 \text{ cm}$

6. Length of shorter side $= \dfrac{73.8 \text{ m}^2}{12.3 \text{ m}} = 6 \text{ m}$

7. To find the area:

   Draw slanted lines to divide the figure into 3 parallelograms, A, B, and C.

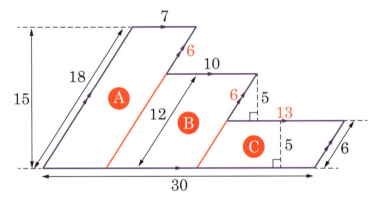

Area of parallelogram $A = 15 \times 7 = 105$
Area of parallelogram $B = 10 \times 10 = 100$
Area of parallelogram $C = 13 \times 5 = 65$
Total Area $= 105 + 100 + 65 = 270$; 270 in^2

Perimeter $= 18 + 7 + 6 + 10 + 6 + 13 + 6 + 30 = 96$; 96 in

Alternatively, the area can be found by dividing the figure with horizontal lines.

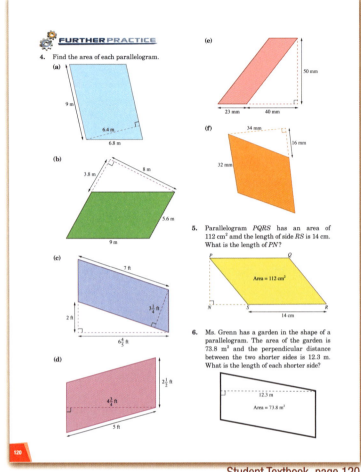

Student Textbook page 120

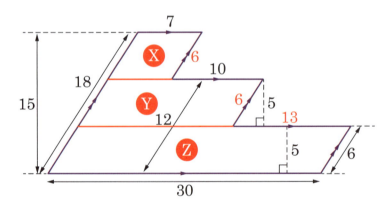

Area of parallelogram $X = 7 \times 5 = 35$
Area of parallelogram $Y = 17 \times 5 = 85$
Area of parallelogram $Z = 30 \times 5 = 150$

Total Area $= 35 + 85 + 150 = 270$; 270 in^2

8. Note: In the textbook's first printing, there is an error in this figure because 38 cannot be the hypotenuse of a right triangle with a leg of 45, and 27 cannot be the hypotenuse of a right triangle with a leg of 32.

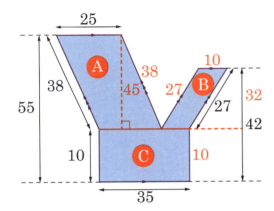

Area of parallelogram $A = 25 \times 45 = 1{,}125$
Area of parallelogram $B = 10 \times 32 = 320$
Area of rectangle $C = 35 \times 10 = 350$
Total area = $1{,}125 + 320 + 350 = 1{,}795$ $1{,}795$ mm^2

Perimeter = $10 + 38 + 25 + 38 + 27 + 10 + 27 + 10 + 35 = 220$, 220 mm

MATH@WORK

9.
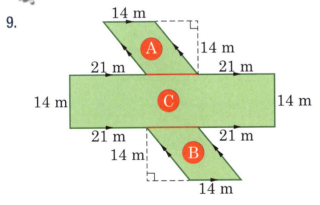

Area of parallelogram $A = 14 \times 14 = 196$
Area of parallelogram $B = 14 \times 14 = 196$
Area of rectangle $C = 14 \times (21 + 14 + 21) = 784$
Total area = $196 + 196 + 784 = 1{,}176$; $1{,}176$ m^2

10. (a) No. AB is the side. The line segment for the height must be parallel to the base.
    (b) For $ABCF$, the base is BC (or AF) and a height is HG. For $CDEF$ the base is CD (or FE) and a height is IJ.

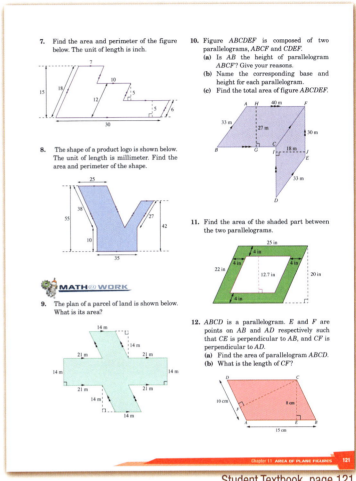

Student Textbook page 121

(c) Area of parallelogram $ABCF = 40$ m $\times 27$ m $= 1{,}080$ m^2
Area of parallelogram $CDEF = 30$ m $\times 18$ m $= 540$ m^2
Total Area of $ABCDEF = 1{,}080$ m$^2 + 540$ m$^2 = 1{,}620$ m^2

11. Area of larger (outside) parallelogram = 25 in $\times$ 20 in = 500 in^2
Area of smaller (inside) parallelogram =
$(25 - 4 - 4)$ in $\times 12.7$ in $\;\;= 17$ in $\times 12.7$ in
$\phantom{(25 - 4 - 4) \text{ in} \times 12.7 \text{ in} \;\;}= 215.9$ in^2
Area of shaded part = 500 in^2 − 215.9 in^2 = 284.1 in^2

12. (a) Area of parallelogram $ABCD = 15$ cm $\times 8$ cm = 120 cm^2
    (b) Length of $CF = \dfrac{120 \text{ cm}^2}{10 \text{ cm}}$ cm = 12 cm

BRAINWORKS

13. (a) $y \times$ height = area of yellow parallelogram

$y \times 2\frac{1}{2} = 3\frac{1}{8}$

$y = 3\frac{1}{8} \div 2\frac{1}{2}$

$y = 1\frac{1}{4}$

(b) Width $= 2\frac{1}{2} = \frac{5}{2}$

Length $= 2\frac{1}{2} + 1\frac{1}{4} = \frac{5}{2} + \frac{5}{4} = \frac{15}{4}$

Width : Length $= \frac{5}{2} : \frac{15}{4} = \frac{10}{4} : \frac{15}{4} = 10 : 15 = 2 : 3$

(c) Area of the flag $= 2\frac{1}{2} \times 3\frac{3}{4}$

$= \frac{5}{2} \times \frac{15}{4}$

$= \frac{75}{8}$

Percentage of the area of yellow part of

the flag $= 3\frac{1}{8} \div \frac{75}{8}$

$= \frac{25}{8} \div \frac{75}{8}$

$= 25 \div 75$

$= 0.3333...$

$\approx 33.3\%$

14. (a) Lines GF and BC are parallel. Thus the perpendicular distance between the lines is the same. If you draw a perpendicular line from BC to GF, it will be the height of all three parallelograms so the height is the same.

(b) The parallelograms all have an equal area because they share a base and the height is the same. This can be shown with dynamic geometry software.

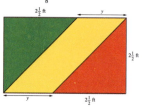
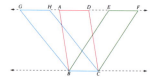
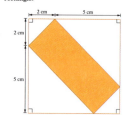
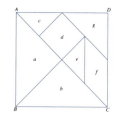

Student Textbook page 122

15. Area of shaded rectangle = Area of square − Area of triangles (A, B, C and D)

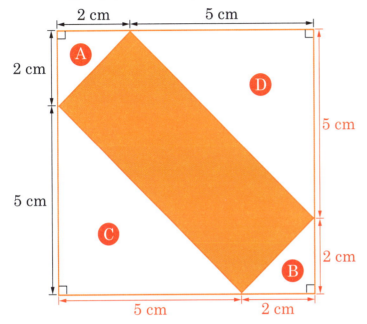

Triangles A and B form a 2 cm × 2 cm square, so the area of both triangles together = 2 cm × 2 cm = 4 cm².

Triangles C and D form a 5 cm × 5 cm square, so the area of both triangles together = 5 cm × 5 cm = 25 cm².

Area of large square = 7 cm × 7 cm = 49 cm²

Area of shaded rectangle = 49 cm² − 4 cm² − 25 cm² = 20 cm²

Note: For triangles C and D, the base angles are the same, which means they are isosceles triangles. Thus, the legs of the triangle must be 5 cm. The length of the side of the square is 7 cm, so the legs of triangle B must be 2 cm each.

16. Note: On the following page, you will find the diagram from this problem that can be copied, cut apart, and distributed to students so they can draw lines or cut.

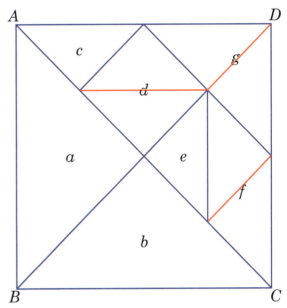

Triangle a and Triangle b are each $\frac{1}{4}$ of the area of the square, so:

Area of $a = \frac{1}{4} \times 1$ m² $= \frac{1}{4}$ m²

Area of $b = \frac{1}{4} \times 1$ m² $= \frac{1}{4}$ m²

8 of shape c will fit on shape a, so $c = \frac{1}{4} \times \frac{1}{4}$ m² $= \frac{1}{16}$ m².

c and e are the same shape, so $e = \frac{1}{16}$ m².

Shapes $d, f,$ and g can each be made with 2 of c, so the areas of $d, f,$ and g are each $2 \times \frac{1}{4}$ m² $= \frac{1}{8}$ m².

Students can also cut the shapes in the upper right half of the tangram into triangles the size of c and see that 8 triangles fit in half of the square, thus 16 fit in the whole square, and determine the fraction of each piece using $\frac{1}{16}$ as the base (e.g., $d, g,$ and f each $= \frac{2}{16} = \frac{1}{8}$, $e = \frac{1}{16}$).

Chapter 11 AREA OF PLANE FIGURES 155

Lesson 4

Objectives:
- Determine the base and height of a triangle.
- Derive the formula for the area of a triangle.

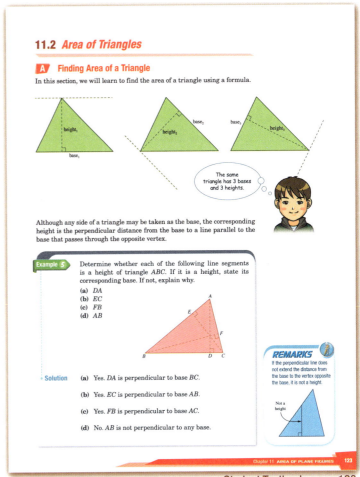

Student Textbook page 123

1. Introduction

Have students draw a triangle on graph paper by drawing 3 points and connecting them with a ruler. (The points should be plotted where the lines of the graph paper intersect.)

Ask, "How is a triangle different from a parallelogram?"
- It has 3 sides.
- It has 3 vertices.
- There are no parallel sides.

Discuss the base and height of a triangle using the top of page 123.

Have students draw a height opposite to each base on their triangle, similar to the top of page 123. Students can use index cards (or protractors) to make sure the height is parallel to the base each time.

Read and discuss the top of page 123 and the REMARKS.

Note: The height of the triangle is the perpendicular distance from the base to the vertex opposite the base. If we draw parallel lines at the base and its opposite vertex, the height can be drawn as a perpendicular line between the parallel lines.

Have students study Example 5, do Try It! 5, and then discuss.

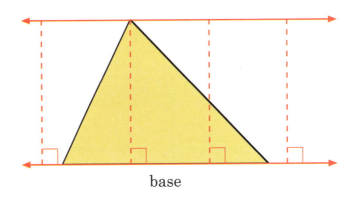
base

2. Development

Have students study Examples 6 – 7 and do Try It! 6 – 7.

Example 6:

- When BC is considered the base, the height AE is drawn outside the base. If we consider AC the base, we can draw the height BD inside the triangle. No sides of this triangle can be considered as the height because none of the sides of the triangle are perpendicular to another side.

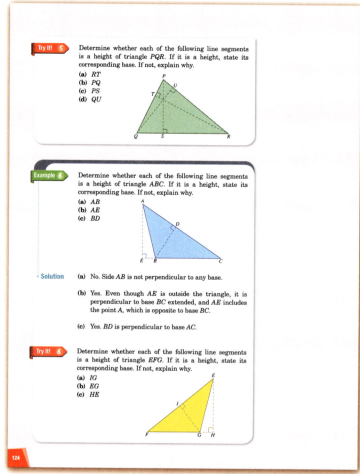

Student Textbook page 124

Try It! 5 Answers

(a) Yes. RT is perpendicular to base PQ.

(b) No. PQ is not perpendicular to any base.

(c) Yes. PS is perpendicular to base QR.

(d) Yes. QU is perpendicular to base PR.

Try It! 6 Answers

(a) Yes. IG is perpendicular to base EF.

(b) No. Side EG is not perpendicular to any base.

(c) Yes. Although HE is outside the triangle, it is perpendicular to base FG extended, and HE includes the point E, which is opposite to base FG.

Have students discuss what the girl is saying on page 125.

Note: For right triangles, a side can be considered as the height because one side is perpendicular to the base.

Try It! 7   Answer

Triangle ABC is a right triangle. Side AC is perpendicular to BC. Thus, AC is the height to base BC.

3. Application

Read the middle of page 125 and remind students how they found the area of the parallelogram by changing it into a rectangle in Class Activity 1.

Give students a piece of blank paper, a ruler, and scissors, and have them do tasks 1 – 3 of Class Activity 2.

Note: Demonstrate how to fold and cut the paper. Depending on how you cut the paper, you will get different triangles that when put together will form different parallelograms.

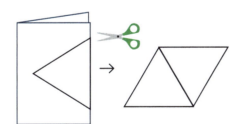

Give students another piece of paper and have them do task #5.

Notes:
- Students should draw and cut a different looking triangle than the first one they made.
- Some students may create right triangles and put them together to make a rectangle, which is fine. Remind them that rectangles are also parallelograms.

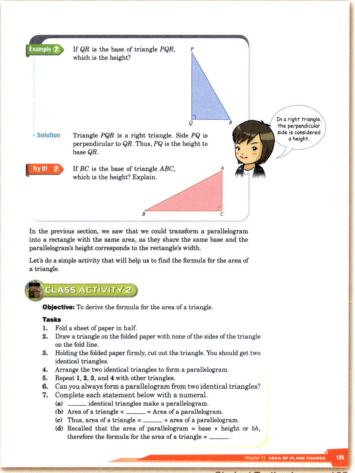

Student Textbook page 125

Have students work together with a partner or in groups to complete tasks 6 – 7 and discuss.

Notes:
- Point out that the base and height of the triangle and the parallelogram are the same.
- For 7 (d), students may say $\frac{1}{2}$ × base × height or base × height ÷ 2. Help them see that they are really the same formula because multiplying by $\frac{1}{2}$ and dividing by 2 give the same result.

Answers for Class Activity 2
6. Yes
7. (a) 2
   (b) $\frac{1}{2}$
   (c) $\frac{1}{2}$
   (d) $\frac{1}{2}$ × base × height (or base × height ÷ 2)

4. **Extension**

Give students a sheet of 1-cm grid paper and ruler, and have them draw triangle *ABC* with a base of 6 cm and a height of 4 cm as shown below.

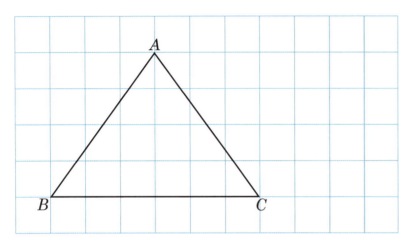

Then, have them draw a rectangle with the same base and height around the triangle as shown below.

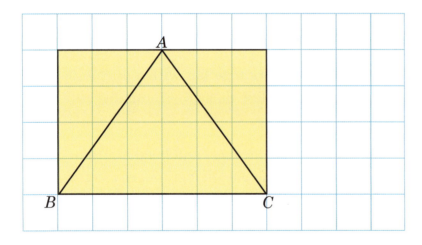

Ask students to find the area of the rectangle and the area of the triangle, compare the areas, and discuss.

Note: The base and height of the triangle and rectangle are the same. The area of the rectangle is 6 cm × 4 cm = 24 cm². The area of the triangle is 12 cm², which is half the area of the rectangle.

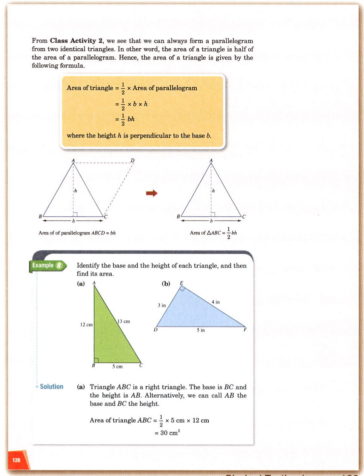

Student Textbook page 126

Have students draw triangle *ABC* again and then draw another identical triangle (upside down) attached to the original triangle as shown below to create a parallelogram.

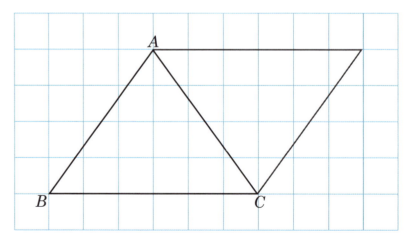

Ask students to find the areas of the parallelogram and the triangle and compare.

160

Note: The base and height of the triangle and the parallelogram are the same. The area of the parallelogram is 6 cm × 4 cm = 24 cm². The area of the triangle 12 cm², which is half the area of the parallelogram. Thus, the area of the triangle is half the area of a parallelogram with the same base and height.

Use the top of page 126 to introduce the formula for finding the area of a triangle and discuss.

Note: The formula for the area of a triangle can also be written as $b \times h \div 2$ or $\frac{bh}{2}$.

5. Conclusion

Summarize the important points of the lesson.
- Any side of a triangle can be considered as the base. The height is always the perpendicular distance from the base to the vertex opposite the base.
- The area of a triangle is half the area of a parallelogram (or rectangle) that has the same base and height.

Lesson 5

Objectives:
- Find the area of triangles.
- Find the base or height of a triangle given the area and the base or height.

1. Introduction

Note: The activities in the **Introduction** and **Development** of this lesson are similar to what students did in Class Activity 2. The purpose is to show that the formula $\frac{1}{2} \times$ base × height works for any triangle, including right triangles and triangles where the height is shown outside the base.

Give students a sheet of blank paper. Have them use a ruler to draw a diagonal line to divide the paper into two right triangles, then cut them out. Have them mark the right angles of each triangle with a square.

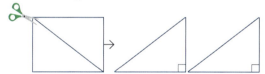

Ask students to identify the perpendicular sides on one of the triangles. Ask, "Where is the base and where is the height?"

Help students see that since two sides of the triangle form perpendicular lines, either of them can be considered the height or the base. The base and height of a right triangle are always adjacent sides.

Then, have them put the triangles together to form a rectangle.

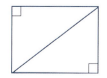

Ask:
- How are the bases and heights of the triangle and the rectangle related? (They are the same.)
- How are the areas of the triangle and the rectangle related? (The area of the rectangle is twice the area of the triangle. The area of the triangle is half the area of the rectangle.)

2. Development

Have students fold a paper (similar to Class Activity 2) and have students draw a triangle similar to the one shown below. Then have them cut out the two triangles.

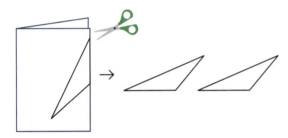

Have students identify the height. They can use an index card or paper to see that the height is perpendicular to the base.

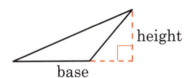

Have students put two triangles together to form a parallelogram.

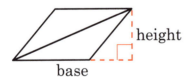

Ask:
- How are the bases and heights of the triangle and the parallelogram related? (They are the same.)
- How are the areas of the triangle and the parallelogram related? (The area of the parallelogram is twice the area of the triangle. The area of the triangle is half the area of the parallelogram.)

Review the formula for the area of a parallelogram on page 126.

Point out that this formula holds true even for right triangles and triangles where the height is shown outside of the base.

3. Application

Have students study Examples 8 – 10 and do Try It! 8 – 10.

Example 8:
- For 8 (a), discuss why the length of side DE cannot be considered as the height. Point out that either of the adjacent perpendicular sides of a right triangle can be considered as the base or height. If we consider EF as the base, then DF is a height. If we consider DF as the base, then EF is a height.
- To calculate the area for 8 (a), students should think about what number is more efficient to multiply by $\frac{1}{2}$ (or divide by 2) first. Since 12 is an even number, it is easier to do $\frac{1}{2} \times 12 \times 5$ (or $12 \div 2 \times 5$) than $\frac{1}{2} \times 5 \times 12$ (or $5 \div 2 \times 12$).
- In 8 (b), point out that even though the triangle is sitting on side QR, QR is not the base. We can consider PR as the base and PQ as the height. Alternatively, we can consider PQ as the base and PR as the height.

Try It! 8 Answers

(a) The base is EF and the height is DF.

Area = $\frac{1}{2} \times 8$ m $\times 6$ m = 24 m^2

(b) The base is PQ and the height is PR.

Area = $\frac{1}{2} \times 7$ cm $\times 24$ cm = 84 cm^2

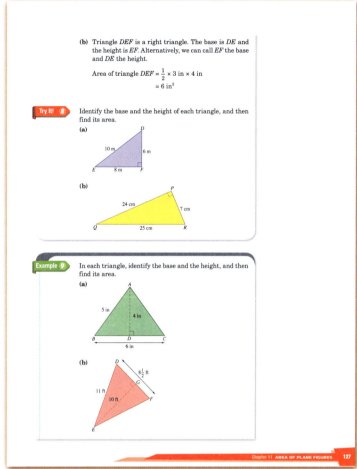

Student Textbook page 127

Example 9:
- These are not right triangles so the length of a side cannot be considered as a height. Have students use an index card to find other possible heights on the triangles.
- For 9 (b), $8\frac{1}{2}$ is changed to an improper fraction. To find the area, it makes more sense to find half of 10 first and then multiply by $8\frac{1}{2}$, or $\frac{17}{2}$. Area is not normally expressed as an improper fraction because it would be harder to get a sense of its size. For example, it is easier to think of the size of $42\frac{1}{2}$ ft^2 than $\frac{85}{2}$ ft^2, so the answer should be expressed as a mixed number in simplest form.

In Try It! 9 (b), the base is a decimal. Students could find half of 5 first and then multiply 2.5 by 5.8.

Try It! 9 Answers

(a) The base is QR. The height is the perpendicular distance from point P to the base. We can add a point S and say that the height is the length of PS.

Area = $\frac{1}{2}$ × 7 cm × 4 cm = 14 cm²

(b) The base is TU. The height is WV.
Area = $\frac{1}{2}$ × 5.8 in × 5 in = 14.5 in²

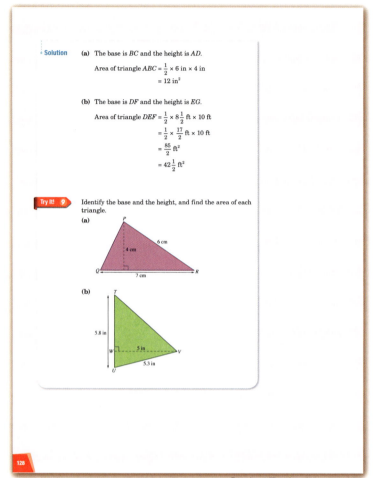

Student Textbook page 128

Example 10:
- In these triangles, the height is drawn outside the base. If students are confused, give them an index card and have them place its corner in the right angle formed by the base and height.
- For Example 10 (a), *WU* is not part of the base. It is a line extended from the base in order to find the height. The same goes for *BD* in Example 10 (b) (See DISCUSS).
- For 10 (b), if students are confused about the base shown as the left side (*AB*), they could turn the book to see *AB* on the bottom.

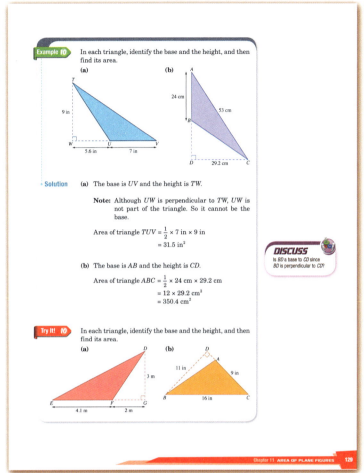

Student Textbook page 129

Try It! 10   Answers

(a) The base is EF and the height is DG.
Area = $\frac{1}{2} \times 4.1$ m $\times 3$ m = 6.15 m^2

(b) The base is AC. The height is BD.
Area = $\frac{1}{2} \times 9$ in $\times 11$ in = 49.5 in^2

4. **Extension**

Have students study Example 11 and then do Try It! 11 and discuss.

Notes:
- In Example 11, we know the area and the base. To find the height, divide the area by $\frac{1}{2}$ of the base.
- $\frac{1}{2} \times$ base $\times$ height = area

  height = $\dfrac{\text{area}}{\frac{1}{2} \times \text{base}}$

- In Try It! 11, we know the area and the height. To find the base, divide the area by $\frac{1}{2}$ of the height.

Try It! 11 Answers

Area of billboard = $\frac{1}{2} \times$ height $\times$ length of edge

480 ft² = $\frac{1}{2} \times$ 25 ft $\times$ length of edge

Length of edge = $\dfrac{480 \text{ ft}^2}{12.5 \text{ ft}}$ = 38.4 ft

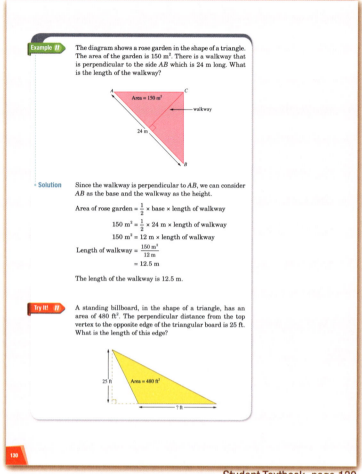

Student Textbook page 130

★ **Workbook: Page 78**

5. **Conclusion**

Summarize the main points of the lesson.
- We can use the formula $\frac{1}{2} \times$ base $\times$ height to find the area of any triangle, even right triangles and triangles where the height is shown outside the base.
- If we know the area of a triangle and the base, we can find the height by dividing the area by half of the base.
- If we know the area of a triangle and the height, we can find the base by dividing the area by half of the height.

Lesson 6

Objective: Find the area of composite plane figures that are composed of parallelograms and triangles.

1. Introduction

Read and discuss page 131. Give students graph paper and rulers and ask them to draw their own composite figures.

Note: Remind students of the composite figures they found the area of in Exercise 11.1 (pages 121 – 122).

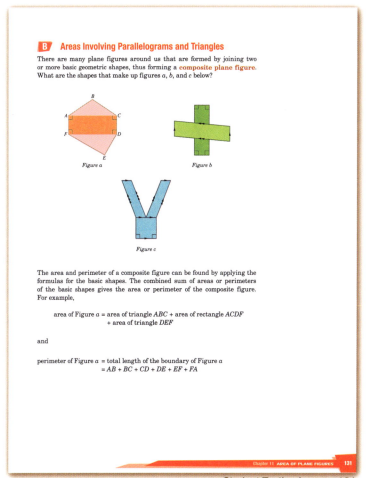

Student Textbook page 131

2. Development

Pose the problem in Example 12 (students should cover the answer with an index card or paper). Ask, "How do you think we can find the area of this shape?" Possible responses:
- Cut it into two (or more) triangles.
- Draw a large rectangle around the shape and subtract out the extra triangles.

Have students try to find the area individually and then share their ideas with a partner or in groups.

Have students share their methods and discuss. Then, compare their methods to Method 1 and Method 2 on pages 132 – 133.

Notes: Students have learned about trapezoids in elementary school but they do not know how to find the areas of trapezoids yet. They can cut the trapezoid into familiar shapes, such as triangles, to find the area.
- Method 1 does not change the area.
- The figure could be also cut into two triangles by drawing a diagonal line from *B* to *D* (see note at bottom of page 132). The calculations will be the same as those shown in Method 1.
- The figure could also be cut into 3 triangles, *ABE*, *AEC*, and *ADC*.

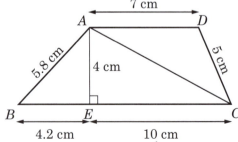

Area of triangle $ABE = \frac{1}{2} \times 4.2 \times 4 = 8.4$

Area of triangle $AEC = \frac{1}{2} \times 10 \times 4 = 20$

Area of triangle $ADC = \frac{1}{2} \times 7 \times 4 = 14$

Area of quadrilateral $ABCD =$
$8.4 \text{ cm}^2 + 20 \text{ cm}^2 + 14 \text{ cm}^2 = 42.4 \text{ cm}^2$

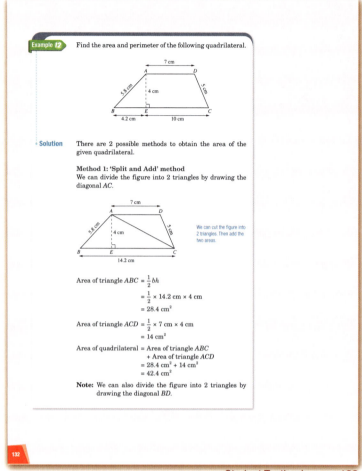

Student Textbook page 132

Method 2 shown on page 133 increases the area, so the extra area needs to be subtracted out to find the area of *ABCD*.

Students already know how to find perimeter by adding the lengths of the sides. To find the perimeter, we do not need to consider the height unless the height is a side, as in right triangles.

3. **Application**

Have students do Try It! 12 on their own.

Have students share and discuss their solutions.

Notes:
- Students can divide the shape into two triangles (Method 1). This method does not involve changing the area.
- Students can find the area of the square and subtract out the areas of the 3 white triangles (Method 2). This involves increasing the area so the extra area has to be subtracted out.

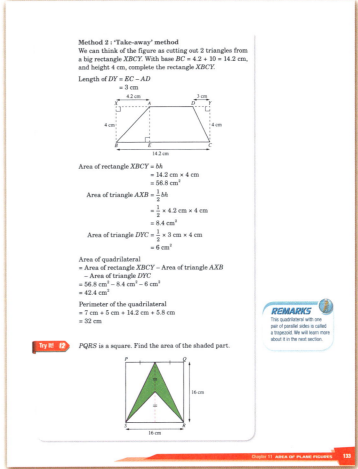

Student Textbook page 133

Try It! 12   Answers

Method 1

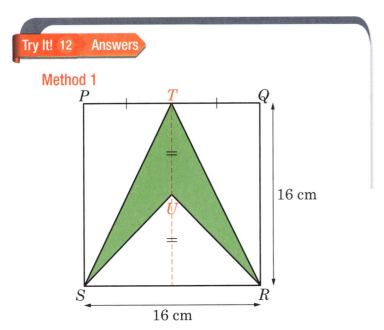

Shaded area = Area of triangle STU + Area of triangle RTU

The base of triangle $STU = \frac{1}{2} \times$ the height of the square $= \frac{1}{2} \times 16$ cm = 8 cm

The height of triangle $STU = \frac{1}{2} \times$ base of the square $= \frac{1}{2} \times 16$ cm = 8 cm

Area of triangle $STU = \frac{1}{2} \times 8$ cm $\times 8$ cm = 32 cm²

Triangle RTU is identical to triangle STU, so the Area of triangle RTU = 32 cm².

Area of shaded part = 32 cm² + 32 cm² = 64 cm²

Method 2

Shaded area = Area of square $PQRS$ − (Area of triangle PTS + Area of triangle QRT + Area of triangle SRU)

Area of square $PQRS$ = 16 cm × 16 cm = 256 cm²

Area of triangle $PTS = \frac{1}{2} \times 8$ cm $\times 16$ cm = 64 cm²

Area of triangle QRT = Area of triangle PTS = 64 cm²

Area of triangle $SRU = \frac{1}{2} \times 16$ cm $\times 8$ cm = 64 cm²

Area of shaded part = 256 cm² − (3 × 64 cm²) = 64 cm²

4. **Extension**

Have students study Example 13 and do Try It! 13.

Example 13:
- Only the bottom side of each square is labeled with the length. Remind students that since they are squares, all the sides are the same length.
- Help students see that the heights of triangle Q and triangle R are the same as the length of one side the middle square. The height of triangle P is the same as the length of a side of the square on the left.

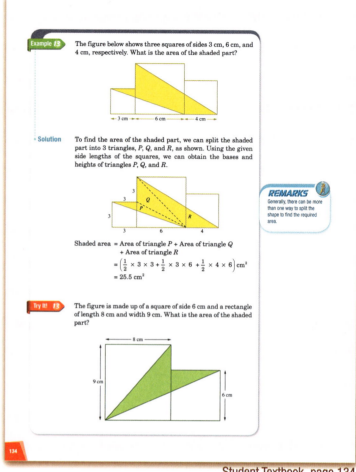

Student Textbook page 134

Try It! 13 Answers

Method 1

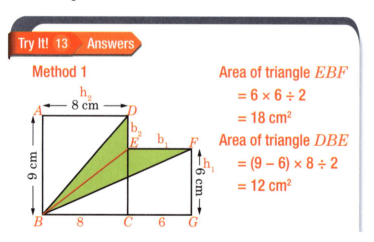

Area of triangle EBF
$= 6 \times 6 \div 2$
$= 18$ cm^2

Area of triangle DBE
$= (9 - 6) \times 8 \div 2$
$= 12$ cm^2

Shaded area $= 18 + 12 = 30$ cm^2

Method 2

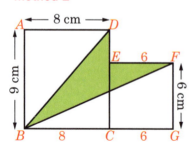

Shaded area = (Area of rectangle $ABCD$ + Area of square $CEFG$) − (Area of triangle ABD + Area of triangle BFG)

Area of rectangle $ABCD = 9$ cm $\times 8$ cm $= 72$ cm^2

Area of square $CEFG = 6$ cm $\times 6$ cm $= 36$ cm^2

Area of triangle $ABD = \frac{1}{2} \times 8$ cm $\times 9$ cm $= 36$ cm^2

Area of triangle $BFG = \frac{1}{2} \times 14$ cm $\times 6$ cm $= 42$ cm^2

Shaded area $= (72$ cm$^2 + 36$ cm$^2) - (36$ cm$^2 + 42$ cm$^2) = 108$ cm$^2 - 78$ cm$^2 = 30$ cm^2

5. **Conclusion**

Summarize the main points of the lesson.
- We can find the area of composite figures by changing them into more familiar shapes that we know how to find the area of.
- If we cut the figure, the area of the figure does not change.
- If we add extra area, we need to subtract out the extra area to find the area of the original shape.

★ **Workbook: Page 84**

Lesson 7

Objective: Consolidate and extend the material covered thus far.

Have students work together with a partner or in groups. Students should try to solve the problems by themselves first, then compare solutions with their partner or group. If they are confused, they can discuss together.

Observe students carefully as they work on the problems. Give help as needed individually or in small groups.

Note: Due to the complexity of the problems, students may use a calculator for some of the problems at the teacher's discretion.

 BASIC PRACTICE

1. (a) Yes. The height is perpendicular to the base and extends to the vertex opposite the base.
   (b) No. The height is perpendicular to the base but it does not extend to the vertex opposite the base.
   (c) No. The height is identified as the base and the base is identified as the height.
   (d) Yes. The height is outside the base but it is perpendicular to the base and it extends to the vertex opposite the base.

2. (a) BC (b) EH
   (c) MK

3. (a) NP (b) VS
   (c) AY

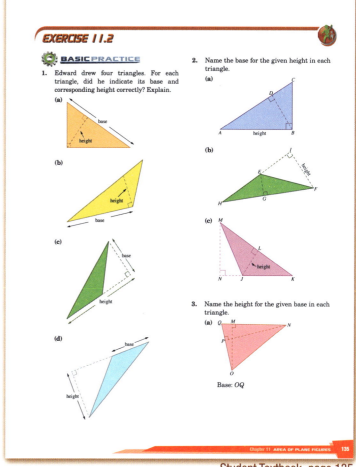

Student Textbook page 135

4. **Question 2**
   (a) AC; BD (b) HF; EG
   (c) JK; MN
   Question 3
   (a) QN; OM (b) RV; TU
   (c) XY; WZ

5. (a) 36 cm²   (b) 37.8 m²
   (c) 90 in²   (d) 18.9 cm²

 FURTHER PRACTICE

6. (a) $\frac{1}{2} \times (68 \text{ cm} + 32 \text{ cm}) \times 93 \text{ cm} = 4{,}650 \text{ cm}^2$

   (b) $\frac{1}{2} \times 45 \text{ mm} \times 24 \text{ mm} = 540 \text{ mm}^2$

   (c) $\frac{1}{2} \times 2.5 \text{ m} \times 5 \text{ m} = 6.25 \text{ m}^2$

   (d) $\frac{1}{2} \times 9 \text{ cm} \times 13 \text{ cm} = 58.5 \text{ cm}^2$

   (e) $\frac{1}{2} \times 10.5 \text{ in} \times 4.8 \text{ in} = 25.2 \text{ in}^2$

   Note: 6 (b) could also be solved by converting mm to cm to find the area.
   $4.5 \times 2.4 \div 2 = 5.4 \text{ cm}^2$

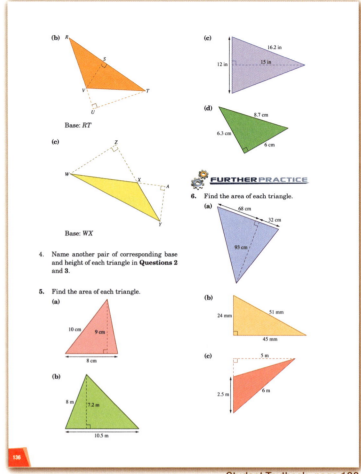

Student Textbook page 136

7. Area of triangle = $\frac{1}{2}$ × 8 cm × length of QR

   16 cm² = 4 cm × length of QR

   Length of QR = $\frac{16 \text{ cm}^2}{4 \text{ cm}}$ = 4 cm

8. Method 1
   Since the triangles are identical, the base of each triangle is 6 cm and the height is $\frac{1}{2}$ × 6 cm = 3 cm.

   Area of each green triangle = $\frac{1}{2}$ × 6 cm × 3 cm = 9 cm²

   Area of both triangles together = 2 × 9 cm² = 18 cm²

   Method 2
   Consider the shape as a large triangle with a base of 6 cm and subtract out the smaller white triangle on the bottom.

   Area = $\left(\frac{1}{2} \times 6 \text{ cm} \times 10 \text{ cm}\right) - \left(\frac{1}{2} \times 6 \text{ cm} \times 4 \text{ cm}\right)$ = 30 cm² − 12 cm² = 18 cm²

   Perimeter = 10.4 cm + 10.4 cm + 5 cm + 5 cm = 30.8 cm

 MATH@WORK

9.

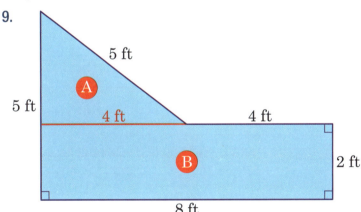

(a) Area of triangle A = $\frac{1}{2}$ × 4 ft × 3 ft = 6 ft²

Area of rectangle B = 8 ft × 2 ft = 16 ft²

Total area = 6 ft² + 16 ft² = 22 ft²

(b) Perimeter = 5 ft + 5 ft + 4 ft + 2 ft + 8 ft = 24 ft

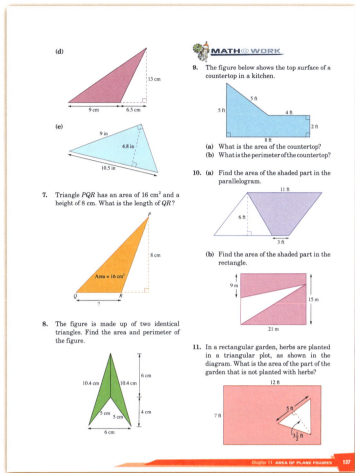

Student Textbook page 137

10. (a)

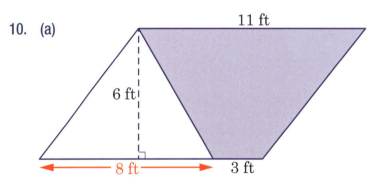

Area of shaded part = Area of parallelogram − Area of triangle

$= (11 \text{ ft} \times 6 \text{ ft}) - \left(\frac{1}{2} \times 8 \text{ ft} \times 6 \text{ ft}\right)$

= 66 ft² − 24 ft²

= 42 ft²

(b)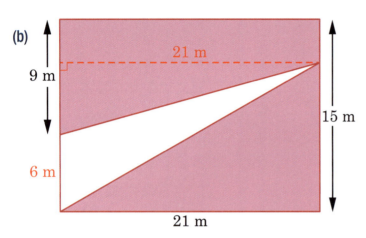

Area of shaded part = Area of rectangle − Area of triangle
= (21 m × 15 m) − ($\frac{1}{2}$ × 6 m × 21 m)
= 315 m² − 63 m²
= 252 m²

11. Area of part not planted with herbs
= Area of rectangle − area of triangle
= (12 ft × 7 ft) − ($\frac{1}{2}$ × 5 ft × $3\frac{1}{2}$ ft)
= 84 ft² − $8\frac{3}{4}$ ft²
= $75\frac{1}{4}$ ft²

12. If we consider BA as the base and JC as the height:
Area of triangle = $\frac{1}{2}$ × 11.2 cm × 15 cm = 84 cm²

If we consider CB as the base and AK as the height:

Area of triangle = $\frac{1}{2}$ × 16 cm × length of AK = 8 cm × length of AK

Length of AK = $\frac{84 \text{ cm}^2}{8 \text{ cm}}$ = 10.5 cm

13. (a) Lines L_1 and L_2 are parallel, thus the perpendicular distance between the lines is the same. The heights are equal, because if we draw a perpendicular line between the base and the top vertex of any of the triangles, the distance is the same.
    (b) The areas are the same because they share the same base and have the same height.

---

12. In triangle ABC, points J and K, on the sides AB and BC respectively, are such that CJ is perpendicular to AB and AK is perpendicular to BC. What is the length of AK?

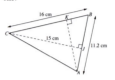

13. The diagram shows three triangles ABC, DBC, and EBC. They share a common base, BC, and their respective opposite vertices A, D and E all lie on a straight line L_1 that is parallel to the base.

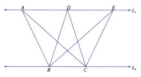

(a) Explain if the heights of the three triangles are equal.
(b) Are the areas of the three triangles equal? Explain your answer.

BRAINWORKS

14. The figure below comprises six square units. If $\frac{7}{n}$ of the entire figure is shaded, what is the value of n?

15. Square P has sides 8 cm and square Q has side 6 cm. The two squares overlap each other. What is the difference in the areas of the two unshaded regions?

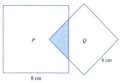

16. Rectangle $ABCD$ is divided into four triangles. Triangles ABE, CBE, and CDE have areas 16 cm², 25 cm², and 24 cm², respectively. What is the area of triangle ADE?

Student Textbook page 138

BRAINWORKS

14.

The entire figure is made up of 6 square units, so the shaded area = $\frac{7}{n}$ × 6.
Shaded area = Area of rectangle − (Area of triangle A + Area of triangle B + Area of triangle C)

$\frac{7}{n}$ × 6 = 6 − ($\frac{1}{2}$ + 2 + $1\frac{1}{2}$)

$\frac{7}{n}$ × 6 = 2

$\frac{7}{n}$ × 6 × n = 2 × n

7 × 6 = 2n

2n = 42

n = 21

15. Overlap area = Area of square P – Area of square Q
    = 64 cm² – 36 cm²
    = 28 cm²

    Note: The difference of the area is the same regardless whether the two squares overlap or not.

    The difference of the area when they are not overlapped:
    $8 \times 8 = 64$ cm²
    $6 \times 6 = 36$ cm²
    So the difference is $64 - 36 = 28$ cm².

    Let's say the area of the overlapping part is 9 cm². The difference of the unshaded regions is:
    $8 \times 8 - 9 = 55$ cm²
    $6 \times 6 - 9 = 27$ cm²
    The difference is $55 - 27 = 28$ cm².

    This is a property of subtraction.

    $A - B = (A - C) - (B - C)$
    $A - B = (A + D) - (B + D)$

    When you add or subtract the same quantity to both numbers, the difference remains the same.

16. Recall problem 13. In that problem, the areas of the three triangles were the same, because the lengths of their bases were the same and the heights (the distance between two parallel lines) were the same. Use this idea to solve this problem.

    The original diagram shown in problem 16 is shown below as Figure 1:

    Figure 1
    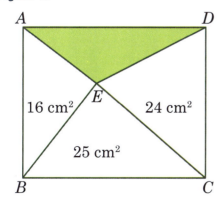

    First, draw a line parallel to line AD that goes through point E. Name the points where this line intersects with lines AB and DC as points P and Q, respectively, so this new line is named line PQ. This is shown below in Figure 2:

    Figure 2
    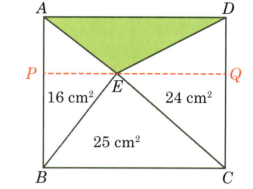

    Move point E of triangle AED to point P. This also moves point E of triangle BEC to point P.

    The area of triangle AED (Figure 2, shown above) and triangle APD (Figure 3, shown below) are the same, because the length of base AD and the height AP are the same in both figures.

    Figure 3
    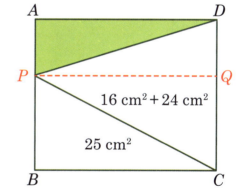

    The area of triangle BEC in Figure 2 and the area of triangle BPC in Figure 3 are the same, because the length of base BC is the same and the height PB is the same.

    Solution continues on next page …

Therefore, the sum of the areas of triangles AEB and CED in Figure 2 is the same as the area of triangle CPD in Figure 3. Because the sum of the areas of triangles AEB and CED is the same as the area of the rectangle $ABCD$ minus the sum of the areas of the triangles AED and BEC, the area of triangle CPD is the same as the area of rectangle $ABCD$ minus the sum of the areas of triangles APD and PBC.

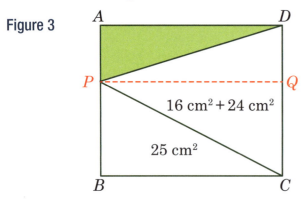

Figure 3

The area of triangle CPD is 16 cm² + 24 cm² = 40 cm², which is half of the area of rectangle $ABCD$.

The other half of the area of rectangle $ABCD$, 40 cm², is made up of the areas of triangles APD and BPC. The area of triangle BPC is 25 cm², so the area of triangle APD is 40 cm² − 25 cm² = 15 cm².

Therefore, the shaded area is 15 cm².

Lesson 8

Objectives:
- Determine the base and height of a trapezoid.
- Derive the formula for the area of a trapezoid and find the area of trapezoids.

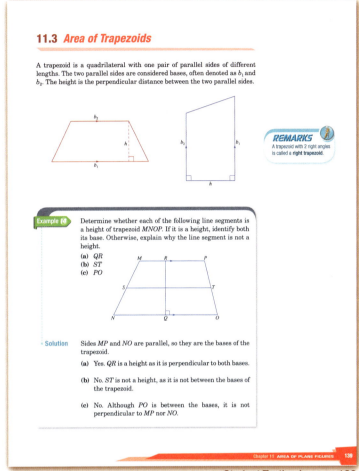

Student Textbook page 139

1. Introduction

Read the top of page 139 and discuss the bases and height of a trapezoid.

A trapezoid is a quadrilateral with one pair of parallel sides. Each parallel side can be considered a base, so a trapezoid has two bases.

Note: On a trapezoid, the height is the perpendicular distance between the top and bottom bases. Since the bases are parallel, many heights can be drawn, even outside the shape.

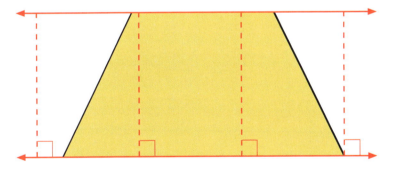

Have students study Examples 14 – 15 and do Try It! 14 – 15.

Example 15:
- These are right trapezoids and may look odd to students who have the image of a trapezoid as always being an isosceles trapezoid (e.g., a pattern block trapezoid). Point out that as long as there is one pair of opposite parallel sides, it is a trapezoid.
- Point out that on a right trapezoid, the side can be considered as a height.

Try It! 14 Answers

(a) No. FG is not a height, as it is not between the bases of the trapezoid.

(b) No. Although DC is between the bases, it is not perpendicular to AD nor BC.

(c) Yes. AE is a height as it is perpendicular to both bases, AD and BC.

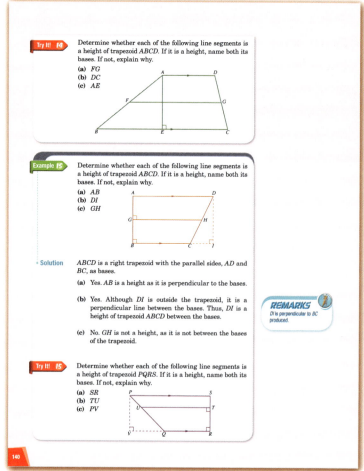

Student Textbook page 140

Try It! 15 Answers

(a) Yes. SR is a height as it is perpendicular to both bases, PS and QR.

(b) No. TU is not a height, as it is not between the bases of the trapezoid.

(c) Yes. Although PV is outside the trapezoid, it is a perpendicular line. Thus, PV is a height of the trapezoid between the bases, PS and QR.

2. Development

Give students 1-cm grid paper and rulers, and have them work together to do Class Activity 3.

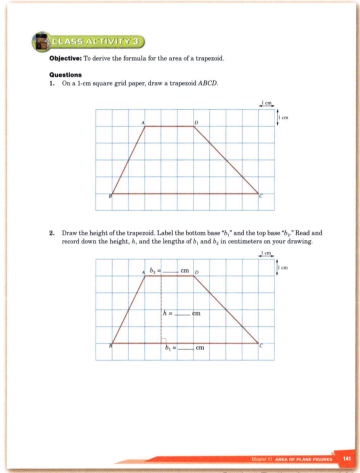
Student Textbook page 141

Have students share and discuss their answers.

Use the bottom of page 142 and page 143 to summarize Class Activity 3.

Notes:
- To find the area of a trapezoid, we take the average of the lengths of the bases and multiply by the height.
- The formula can also be written:

$(b_1 + b_2) \div 2 \times h$ or $\dfrac{b_1 + b_2}{2} \times h$

Answers for Class Activity 3

3. (a) Parallelogram *ABEF*

   (b) 9 cm + 3 cm = 12 cm

   (c) 9 cm + 3 cm = 12 cm

   (d) The sum of the bases of the trapezoids is the same as the base of parallelogram *ABEF*.

4. (a) **Note:** Students can count the squares and half-squares of the figures or use calculation.

   Area of figure *ABEF* =
   12 cm × 4 cm = 48 cm²

   Area of trapezoid *ABCD* =
   $\frac{1}{2}$ × 48 cm² = 24 cm² (or 48 cm² ÷ 2 = 24 cm²)

   (b) The area of trapezoid *ABCD* is half the area of figure *ABEF*.
   The trapezoid is half the size of the parallelogram.

5. $\frac{1}{2}$ × (the top base + the bottom base) × height
   Or,
   (the bottom base + the top base) × height ÷ 2

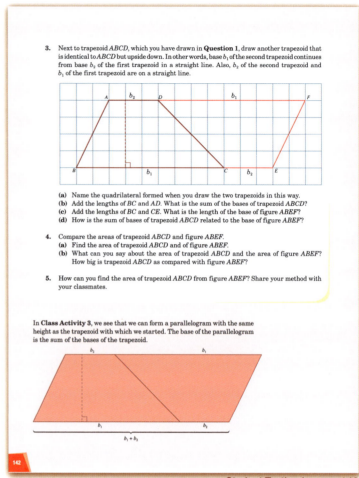

Student Textbook page 142

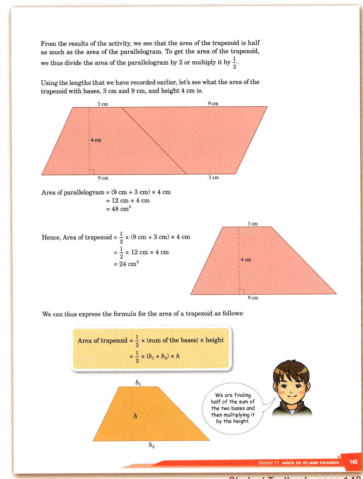

Student Textbook page 143

3. Application

Have students study Examples 16 – 17 and do Try It! 16 – 17.

Example 16:
- Students should think about the order they want to multiply to make the calculation easier. In Example 16, it is easier to multiply $\frac{1}{2}$ by 4 first and then multiply 13 by $\frac{1}{2}$.
- In Try It! 16 (b), remind students that is it a right trapezoid and the perpendicular side can be considered as the height.

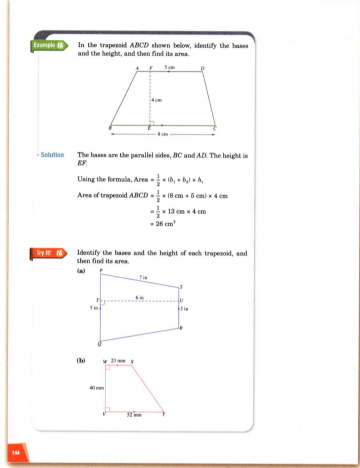

Student Textbook page 144

Try It! 16   Answers

(a) The bases are the parallel sides PQ and SR.
The height is TU.
Area of trapezoid $PQRS$
$= \frac{1}{2} \times (5 \text{ in} + 3 \text{ in}) \times 6 \text{ in}$
$= \frac{1}{2} \times 8 \text{ in} \times 6 \text{ in}$
$= 24 \text{ in}^2$

(b) The bases are the parallel sides VY and WX. The height is side WV because it is perpendicular to the bases.
Area of trapezoid $WXYZ$
$= \frac{1}{2} \times (52 \text{ mm} + 23 \text{ mm}) \times 40 \text{ mm}$
$= \frac{1}{2} \times 75 \text{ mm} \times 40 \text{ mm}$
$= 1{,}500 \text{ mm}^2$

This could be solved by converting mm to cm.

Area of trapezoid $WXYZ = \frac{1}{2} \times (5.2 + 2.3) \times 4$
$= 15 \text{ cm}^2$

In Example 17 and Try It! 17 (a), the height is shown outside the base. Remind students that *HG* is not part of the base in Example 17, and *JN* is not part of the base in Example 17 (b).

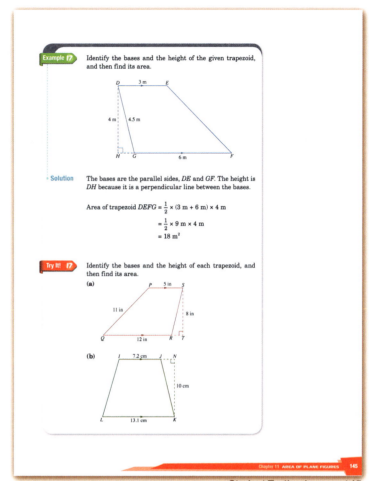

Student Textbook page 145

Try It! 17 Answers

(a) The bases are the parallel sides QR and PS.
The height is ST.
Area of trapezoid $PQRS$
$= \frac{1}{2} \times (12 \text{ in} + 5 \text{ in}) \times 8 \text{ in}$
$= \frac{1}{2} \times 17 \text{ in} \times 8 \text{ in}$
$= 68 \text{ in}^2$

(b) The bases are the parallel sides LK and IJ.
The height is NK.
Area of trapezoid $IJKL$
$= \frac{1}{2} \times (13.1 \text{ cm} + 7.2 \text{ cm}) \times 10 \text{ cm}$
$= \frac{1}{2} \times 20.3 \text{ cm} \times 10 \text{ cm}$
$= 101.5 \text{ cm}^2$

4. Extension

Have students study Example 18 and do Try It! 18.

Notes:
- In Example 18, we are finding a missing base. We can substitute a variable for the missing base and solve algebraically.
- In Try It! 18, we are finding a missing height. To find the height, divide the area by half of the sum of the bases.

$$\frac{1}{2} \times \text{(sum of bases)} \times \text{height} = \text{Area}$$

$$\text{Height} = \frac{\text{Area}}{\frac{1}{2} \times \text{sum of bases}}$$

Try It! 18   Answer

Area of bulletin board = $\frac{1}{2} \times$ (sum of bases) × height

32 ft² = $\frac{1}{2} \times$ 12.8 ft × height

32 ft² = 6.4 ft × height

Height = $\frac{32 \text{ ft}^2}{6.4 \text{ ft}}$ = 5 ft

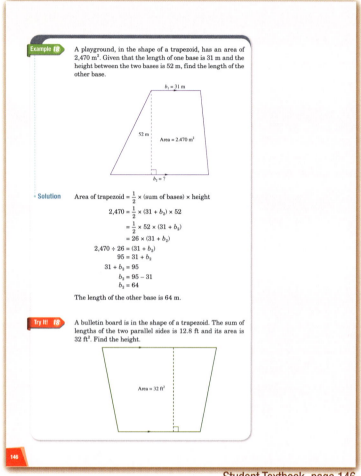

Student Textbook page 146

★ **Workbook: Page 88**

5. Conclusion

Summarize the important points of the lesson.
- A trapezoid has two bases which are opposite parallel sides.
- The height of a trapezoid is the perpendicular distance between the top and bottom bases.
- The area of a trapezoid is half the area of a parallelogram with the same height and a base that is equal to the sum of the bases of the trapezoid.
- To find the area of a trapezoid, we multiply half of the sum of the bases by the height (i.e. the average length of the two bases × height).

Lesson 9

Objective: Consolidate and extend the material covered thus far.

Have students work together with a partner or in groups. Students should try to solve the problems by themselves first, then compare solutions with their partner or group. If they are confused, they can discuss together.

Observe students carefully as they work on the problems. Give help as needed individually or in small groups.

Note: Due to the complexity of the problems, students may use a calculator for some problems at the teacher's discretion.

BASIC PRACTICE

1. (a) (i) The bases are AD and BC.
      (ii) She identified the height correctly because it is a perpendicular line between the bases.
   (b) (i) The bases are EG and IH.
      (ii) She identified the height incorrectly because she identified a side that is not a perpendicular line. HJ is the correct height.
   (c) (i) The bases are KL and MN.
      (ii) She identified the height incorrectly because the perpendicular line is not in between the bases. KN is the correct height.
   (d) (i) The bases are OP and RQ.
      (ii) She identified the height correctly because it is a perpendicular line outside one of the bases that is the perpendicular distance to the other base.

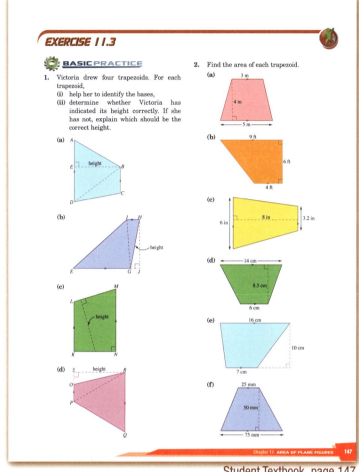

Student Textbook page 147

2. (a) 16 m²   (b) 39 ft²
   (c) 36.8 in²   (d) 85 cm²
   (e) 115 cm²   (f) 2,500 mm²

FURTHER PRACTICE

3. (a) $\frac{1}{2} \times (8.2 \text{ m} + 7.8 \text{ m}) \times 6.5 \text{ m} = 52 \text{ m}^2$

   (b) $\frac{1}{2} \times (4\frac{3}{4} \text{ ft} + 2\frac{1}{2} \text{ ft}) \times 4 \text{ ft} = 14\frac{1}{2} \text{ ft}^2$

   (c) $\frac{1}{2} \times (36 \text{ cm} + 28 \text{ cm}) \times 25 \text{ cm} = 800 \text{ cm}^2$

   (d) $\frac{1}{2} \times (46 \text{ mm} + 17 \text{ mm}) \times 32 \text{ mm} = 1{,}008 \text{ mm}^2$

4. (a) Area of $ABCD = \frac{1}{2} \times (6 \text{ in} + 13 \text{ in}) \times$ length of AB

   $76 \text{ in}^2 = 9.5 \text{ in} \times$ length of AB

   Length of $AB = \frac{76 \text{ in}^2}{9.5} = 8 \text{ in}$

   (b) Area of triangle $BCD = \frac{1}{2} \times 13 \text{ in} \times 8 \text{ in} = 52 \text{ in}^2$

5. (a) Area of $PQRS = \frac{1}{2} \times (16 + b_2) \times 10$

   $105 = \frac{1}{2} \times 10 \times (16 + b_2)$

   $105 \div 5 = 16 + b_2$

   $21 = (16 + b_2)$

   $b_2 = 21 - 16$

   $b_2 = 5$

   Length of $PS = 5$ cm

   (b) Area of triangle $RSP = \frac{1}{2} \times 5 \text{ cm} \times 10 \text{ cm} = 25 \text{ cm}^2$

6. (a) $GM = \frac{1}{2} \times (6 \text{ in} + 12 \text{ in}) = 9 \text{ in}$

   (b) $198 \text{ in}^2 = 18 \text{ in} \times h$

   $h = \frac{198 \text{ in}^2}{18 \text{ in}}$

   $= 11 \text{ in}$

   (c) $\frac{1}{2} \times (6 \text{ in} + 9 \text{ in}) \times 11 \text{ in} = 82.5 \text{ in}^2$

 MATH@WORK

7. Area of rectangle = $42 \text{ yd} \times 20 \text{ yd} = 840 \text{ yd}^2$

   Area of trapezoid = $\frac{1}{2} \times (42 \text{ yd} + 30 \text{ yd}) \times 20 \text{ yd} = 720 \text{ yd}^2$

   Fraction of area not for pets = $\frac{720}{840} = \frac{6}{7}$

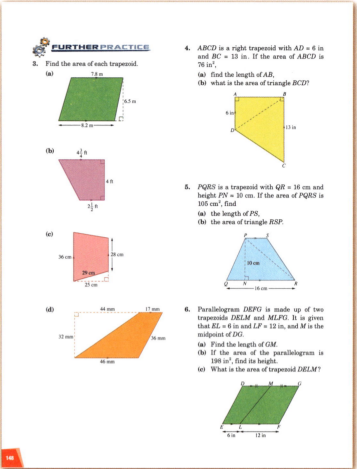

Student Textbook page 148

8. (a) Area of rectangle = $32 \text{ in} \times 25 \text{ in} = 800 \text{ in}^2$

   Area of trapezoid = $\frac{1}{2} \times (32 \text{ in} + 18 \text{ in}) \times 25 \text{ in} = 625 \text{ in}^2$

   Combined area of triangles = $800 \text{ in}^2 - 625 \text{ in}^2 = 175 \text{ in}^2$

   Alternatively: Move the top of the trapezoid to the right edge of the rectangle to create one large blue triangle with base = $(32 \text{ in} - 18 \text{ in})$ and height = 25 in

   $\frac{1}{2} \times (32 \text{ in} - 18 \text{ in}) \times 25 \text{ in} = 175 \text{ in}^2$

   (b) $\frac{625}{800} \times 100\% \approx 78.1\%$

   Note: the actual answer is 78.125% but it can be rounded off to 78.1% or 78%.

9. Note: It may be helpful to make a sketch.

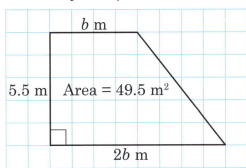

(a) $49.5 = \frac{1}{2} \times (2b + b) \times 5.5$

$= \frac{1}{2} \times 5.5 \times 3b$

$49.5 \div 2.75 = 3b$

$18 = 3b$

$b = 6$

(b) 2×6 m $= 12$ m

10. (a) $1{,}080 = \frac{1}{2} \times (50 + 40) \times h$

$= \frac{1}{2} \times 90 \times h$

$= 45 \times h$

$h = 1{,}080 \div 45$

$= 24$

The distance between the rungs is 24 cm.

(b) Area of $BCRQ = \frac{1}{2} \times (60$ cm $+ 50$ cm$) \times 24$ cm

$= 1{,}320$ cm^2

11. (a) $\frac{2}{3} = \frac{28}{x}$

$\frac{2}{3} \times x = \frac{28}{x} \times x$

$\frac{2}{3}x = 28$

$x = 28 \times \frac{3}{2}$

$x = 42$

$BC = 42$ in

$OM = 42$ in $\div 2 = 21$ in

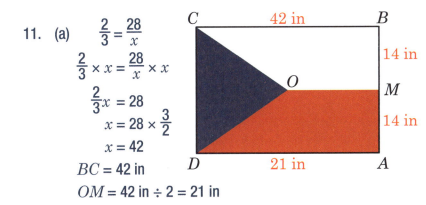

Student Textbook page 149

(b) The white part is $\frac{3}{8}$ of the area of the rectangle.

$\frac{3}{8} \times 42 \times (14 + 14) = 441$ in^2

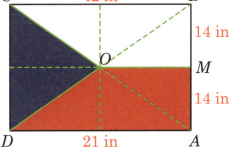

(c) The red area is $\frac{3}{8}$ of the area of the flag.

$\frac{3}{8} \times 100\% = 37.5\%$

12.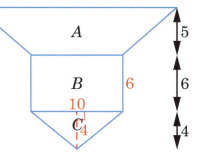

(a) Area of rectangle = 10 cm × 6 cm = 60 cm²

Area of triangle = $\frac{1}{2}$ × 10 cm × 4 cm = 20 cm²

Combined area of rectangle and triangle =
60 cm² + 20 cm² = 80 cm²

(b) 80 cm² = $\frac{1}{2}$ × (10 cm + longer base) × 5 cm
= 2.5 cm × (10 cm + longer base)

80 cm² ÷ 2.5 cm = 10 cm + longer base

32 cm = 10 cm + longer base

Longer base = 32 cm − 10 cm
= 22 cm

BRAINWORKS

13. ABE and AED share the same height, and BCE and CDE share the same height.

Since ABE and BCE share the same base, and AED and CDE share the same base, the ratio of the area of ABE to the area of AED = ratio of the area of BEC to the area of CDE.

$\frac{72}{36} = \frac{\text{Area of } BCE}{30}$

$\frac{72}{36} \times 30 = \text{Area of } BCE$

Area of BCE = 60 cm²

72 cm² + 36 cm² + 30 cm² + 60 cm² = 198 cm²

Area of $ABCD$ = 198 cm²

14. 4 cm + 4 cm + 14 cm + 4 cm + 4 cm = 30 cm

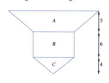
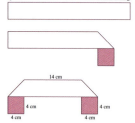
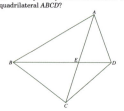
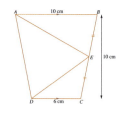

Student Textbook page 150

15. Area of trapezoid $ABCD$ =
$\frac{1}{2}$ × (10 cm + 6 cm) × 10 cm = 80 cm²

Area of triangle ABE = $\frac{1}{2}$ × 10 cm × 5 cm = 25 cm²

Area of triangle BCE = $\frac{1}{2}$ × 6 cm × 5 cm = 15 cm²

Area of triangle ADE = 80 cm² − (25 cm² + 15 cm²) = 40 cm²

Lesson 10

Objective:
- Summarize and reflect on important ideas learned in this chapter, and solve a non-routine problem.

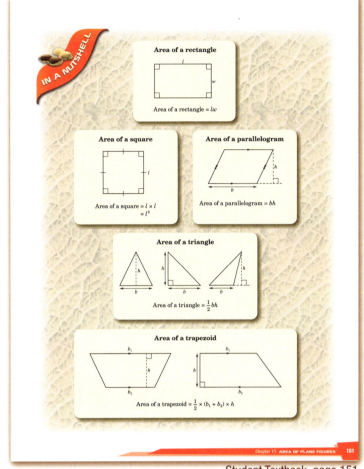

Student Textbook page 151

Note: This lesson could be done in class or assigned for students to do independently at home or at school.

1. In a Nutshell

Use this page to summarize the important ideas learned in this chapter.

Give examples where needed.

2. Write in Your Journal

Have students do the writing activity and share their answers. Answers will vary.

Notes:
- To explore this idea, have students draw a trapezoid on graph paper, draw a rectangle around it, and subtract the areas of the two triangles from the area of the rectangle.
- They can also draw a trapezoid, draw a diagonal to cut it into two triangles, and find the sum of the areas of the triangles. The sum of the areas of the triangles = $\frac{1}{2} \times b_1 \times \text{height} + \frac{1}{2} \times b_2 \times \text{height} = \frac{1}{2} \times (b_1 + b_2) \times \text{height}$.

3. Extend Your Learning Curve

This activity can be completed in class or done as an independent assignment.

(a) You will get a parallelogram. If the trapezoid is an isosceles trapezoid, you will get a rhombus. Depending on the shape of the trapezoid, the parallelogram could be a rectangle or rhombus.

(b) You will get a parallelogram.

(c) A square, a rectangle, or an isosceles trapezoid.

Notes:
- Problem (c) is an application of the midsegment theorem in geometry. "In a triangle, the segment joining the midpoints of any two sides will be parallel to the third side and half its length." A parallelogram formed in a quadrilateral when midpoints of each sides are connected is called a Varignon parallelogram. The theorem is called Varignon's Theorem. Students usually

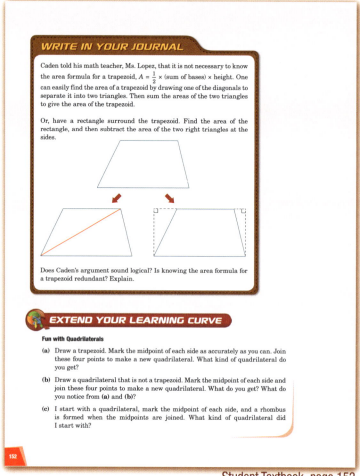

Student Textbook page 152

learn this in high school geometry class in the context of learning proof. This is an investigation activity, so students need to draw many different kinds of quadrilaterals and generalize the conclusion inductively.

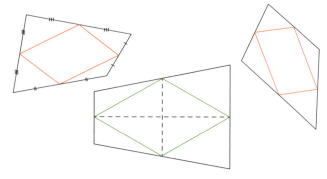

- For (c), a rhombus is formed when a quadrilateral has equal length diagonals, which means it could be a rectangle (or a square), or an isosceles trapezoid.

Chapter 12: Volume and Surface Area of Solids

Lesson	Objectives	Class Periods	Textbook & Workbook	Teacher's Guide Page	Additional Materials Needed
1	• Understand the properties of right rectangular prisms. • Find the volume of cubes and cuboids. • Find the volume of a cube with fractional edge lengths.	1	TB: 153–158	195	Examples of prisms, 1-cm grid paper, scissors, tape, 1-cm cubes
2	• Find the volume of right rectangular prisms with fractional edge lengths.	1	TB: 159–162	201	2 boxes (or other cuboids) of different sizes, *Optional:* two different sized clear cube shaped beakers
3	• Find the volume of compound solid figures composed of cuboids. • Find the volume of liquids in a container.	1	TB: 163–166 WB: 93–100 WB: 101–109	206	
4	• Consolidate and extend the material covered thus far.	1	TB: 167–169	211	
5	• Identify various types of prisms. • Find the surface area of right rectangular prisms.	1	TB: 170–175 WB: 110–118	216	Examples of prisms, 1-cm grid paper, scissors
6	• Find the surface area of triangular prisms.	1	TB: 175–179 WB: 119–121	221	1-cm grid paper, scissors
7	• Convert between metric units of area and volume. • Solve problems involving area, volume, and surface area that require measurement conversions.	1	TB: 180–183 WB: 122–125	226	6 meter sticks, 1-cm cubes
8	• Consolidate and extend the material covered thus far.	1	TB: 184–186	234	

Continues on next page.

Chapter 12: Volume and Surface Area of Solids

Lesson	Objectives	Class Periods	Textbook & Workbook	Teacher's Guide Page	Additional Materials Needed
9	• Summarize and reflect on important ideas learned in this chapter, and solve a non-routine problem.	1	TB: 187–188	237	Nets of triangular pyramids and triangular prisms, small square papers
10	• Solve complex and non-routine problems involving solid figures.	1	TB: 189–192	239	

A **prism** is a solid figure that has two parallel, identical end faces, also called bases. Prisms are named for the shape of the faces at each end. In this chapter, students will find the volume and surface area of right rectangular and right triangular prisms. A **right prism** is a prism whose lateral faces are perpendicular to the bases. A right rectangular prism is also called a **cuboid**.

In elementary grades, students learned about capacity and liquid volume, that 1 mL = 1 cm³, and that 1 L = 1,000 mL = 1,000 cm³. In Grade 5, they derived the formula for the volume of a cuboid with sides that measured an exact number of centimeters by filling with 1-cm cubes.

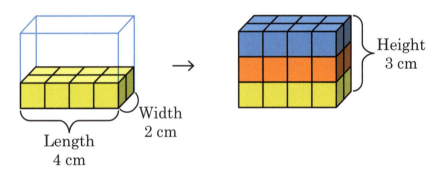

The product of the length and width gives the number of unit cubes that will cover the bottom layer. The height tells the number of layers needed to fill the cuboid. Since length × width is the same as finding the area of the base, we can find the volume of the cuboid using the formula Volume of a cuboid = Area of base × height.

Knowing this formula, **Area of base × height**, will help prepare students to find the volume of different kinds of prisms in future grades as it can be used to find the volume of any prism, regardless of the shape of its base. It can also be used to find the volume of a cylinder.

In Grade 5, students also found the volumes of compound solids composed of cuboids using the methods shown on the following page.

Chapter 12: Volume and Surface Area of Solids

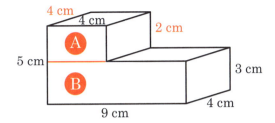

Cut the solid either horizontally or vertically into two cuboids. For example,

Volume of cuboid A = 4 cm × 4 cm × 2 cm = 32 cm³
Volume of cuboid B = 9 cm × 4 cm × 3 cm = 108 cm³
Volume of solid figure = 32 cm³ + 108 cm³ = 140 cm³

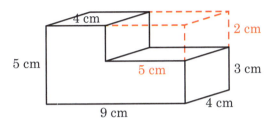

Add volume to create a cuboid and then subtract out the extra volume.

Volume of large cuboid = 9 cm × 4 cm × 5 cm = 180 cm³
Volume of small cuboid = 5 cm × 4 cm × 2 cm = 40 cm³
Volume of solid figure = 180 cm³ − 40 cm³ = 140 cm³

These methods are similar to the methods students used in **Dimensions Math® 6** Chapter 11 to find the areas of complex shapes.

In this chapter, students will find the volumes of solids with fractional edge lengths. To help students understand this conceptually, it is important to help them understand the relationship between the volume of a unit cube and the volume of a cube where the length of the edge is a fraction.

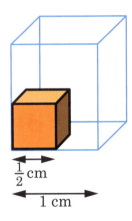

Two $\frac{1}{2}$-cm cubes will fit along each edge.

It takes 2 × 2 × 2, or eight, $\frac{1}{2}$-cm cubes to fill the 1-cm cube.

Thus, the volume of a $\frac{1}{2}$-cm cube is $\frac{1}{8}$ of the volume of a 1-cm cube.

The volume of a 1-cm cube = 1 cm³

Thus, the volume of a $\frac{1}{2}$-cm cube = $\frac{1}{8}$ cm³.

If we multiply length × width × height, we get:

$\frac{1}{2}$ cm × $\frac{1}{2}$ cm × $\frac{1}{2}$ cm = $\frac{1}{8}$ cm³

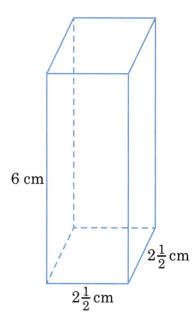

Five $\frac{1}{2}$-cm cubes will fit along the length.

Five $\frac{1}{2}$-cm cubes will fit along the width.

Twelve $\frac{1}{2}$-cm cubes will fit along the height.

It takes $5 \times 5 \times 12$ or 300 $\frac{1}{2}$-cm cubes fill up the prism.

Volume of one $\frac{1}{2}$-cm cube = $\frac{1}{8}$ cm³

Volume of 300 $\frac{1}{2}$-cm cubes = $300 \times \frac{1}{8}$ cm³ = $37\frac{1}{2}$ cm³

If we multiply length × width × height, we get:

$2\frac{1}{2}$ cm × $2\frac{1}{2}$ cm × 6 cm = $37\frac{1}{2}$ cm³

In this chapter, students will also learn to find the surface areas of rectangular and triangular prisms. Creating a net of a prism helps us see the shapes of the faces used to create the 3D figure.

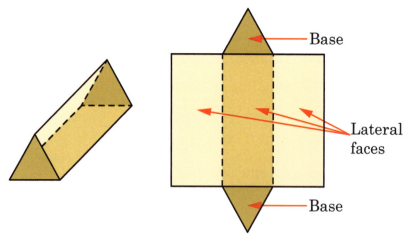

The net allows us to see the elements that make up the surface of the 3D figure, how the edges of each face are connected to the edges of other faces, and determine the lengths of these connected edges. Students who struggle with perspective drawing often struggle with surface area and volume. In part, they are struggling with understanding how edges come together to makes points and faces form together to make edges on surfaces. For example, often students are surprised to realize triangular prisms have rectangles for sides. Significant practice forming nets from 3D shapes, and forming 3D shapes from nets, is necessary for students to recognize how 2D shapes build up to make 3D shapes.

Calculating the surface area, then, is simply a matter of calculating the areas of the 2D shapes in the net. To find the surface area of a prism, we add the areas of the bases (combined base area) and areas of the faces (lateral surface area).

Lesson 1

Objectives:

- Understand the properties of right rectangular prisms.
- Find the volume of cubes and cuboids.
- Find the volume of a cube with fractional edge lengths.

1. Introduction

Discuss page 153.
State the learning goals of the chapter, LET'S LEARN TO …

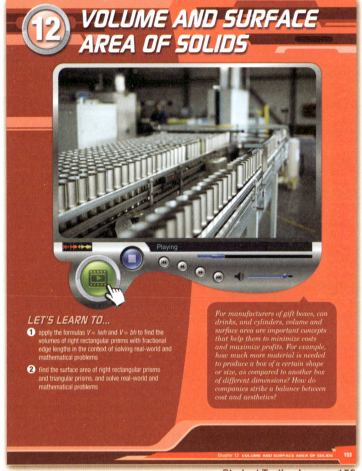

Student Textbook page 153

Show students some prisms; e.g., rectangular prism, triangular prism. (See page 170 for examples.)

Read and discuss page 154.

Have students identify the faces, bases, vertices and edges on a real-life cuboid such as a box or a die.

Notes:

- A cuboid is a solid that is shaped like a box. It is also known as a right rectangular prism. The bases are congruent rectangles. The lateral faces are all rectangles that are perpendicular to the bases (thus, the meaning of "right" prism).
- A net is the 2D representation of a 3D shape made by cutting as few edges as possible to get the faces of the 3D shape to lie flat. It may be helpful to actually cut the edges of a box to illustrate this.
- Help students see that any face of a cuboid can be considered the base by placing a cuboid on its different faces.

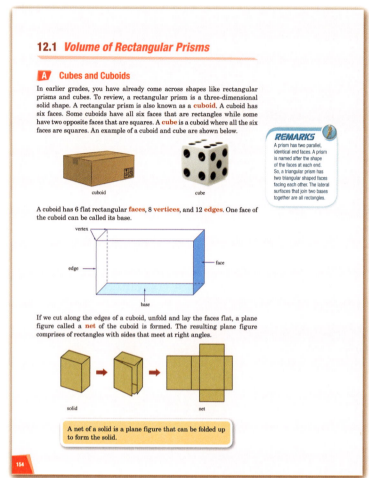

Student Textbook page 154

2. Development

Give groups or pairs of students some 1-cm cubes and have them build a cube and a cuboid.

Ask the students to find the volumes of each by counting the cubes and using the two formulas length × width × height and Area of base × height.

Read and discuss page 155, including the REMARKS and RECALL.

Notes:
- We measure volume in cubic units. 1 cm^3 is read as "1 cubic centimeter" (not "1 centimeter cubed").
- For a cuboid, any side can be considered the base, so to find the volume of the cuboid they can multiply the area of any face by the length of an edge that is not part of that face (e.g., Area of front face × width).
- Since the lengths of the edges of a cube are all the same, we only need to know the length of one edge to find the volume. We can use the formula **edge × edge × edge** or **edge^3**.
- The length tells the number of 1-cm cubes that fit along the front of the base, and the width tells the number of 1-cm cubes that fit along the side of the base. Thus, length × width tells the number of cubes in one layer, which is the same as the area of the base. The height tells the number of layers.

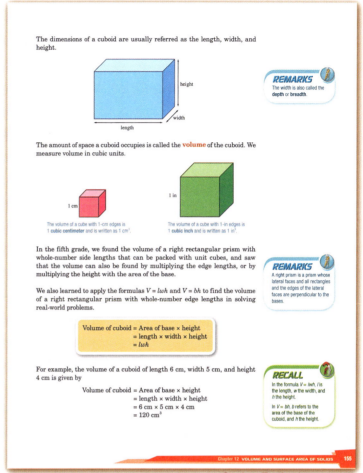

Student Textbook page 155

3. Application

Discuss the top of page 156.

Have students work in groups or with a partner to do Class Activity 1.

Notes:
- Since the denominator divides the unit into equal parts, the denominator tells us how many cubes with a fractional side length fit along each edge of a 1-unit cube. For example, for a $\frac{1}{2}$-unit cube, 2 cubes will fit along each edge; for a $\frac{1}{3}$-unit cube, 3 cubes will fit across each edge; etc.

Answers for Class Activity 1

1. (a) 2 cubes can fit across the length, 2 across the width, and 2 in the height.

    (b) $2 \times 2 \times 2 = 8$; 8 cubes

2. (a) 3 cubes can fit across the length, 3 across the width, and 3 in the height.

    (b) $3 \times 3 \times 3 = 27$; 27 cubes

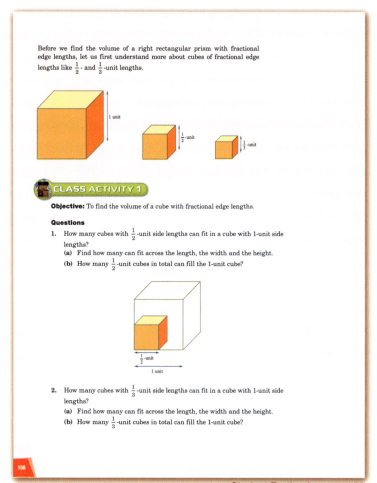

Student Textbook page 156

3. 8

   $\frac{1}{8}$

   27

   $\frac{1}{27}$

- Since it takes eight $\frac{1}{2}$-unit cubes to fill a 1-unit cube, the volume of the $\frac{1}{2}$-unit cube is $\frac{1}{8}$ of the volume of a 1-unit cube.

- Since it takes twenty-seven $\frac{1}{3}$-unit cubes to fill a 1-unit cube, the volume of the $\frac{1}{3}$-unit cube is $\frac{1}{27}$ of the volume of a 1-unit cube.

4. $\frac{1}{8}$

   $\frac{1}{27}$

- The purpose of Question 4 is to show that the formula length × width × height (or edge × edge × edge) will also work when the length of an edge is a fraction.

5. The volume of a cube with fractional edge lengths is the same as the product of the edge lengths.

Use the bottom of page 157 to summarize **Class Activity 1**.

---

3. Compare the number of cubes you obtained in **Questions 1** and **2** with the volume of a 1-unit cube. What does this mean about the volume of a $\frac{1}{2}$-unit cube and a $\frac{1}{3}$-unit cube?

   Complete the following statements:

   Volume of a 1-unit cube = 1 cubic unit

   From **1(b)**, there are ☐ cubes with $\frac{1}{2}$-unit side lengths in 1 cubic unit.

   Hence, the volume of one cube with $\frac{1}{2}$-unit side lengths is ☐ cubic unit.

   From **2(b)**, there are ☐ cubes with $\frac{1}{3}$-unit side lengths in 1 cubic unit.

   Hence, the volume of one cube with $\frac{1}{3}$-unit side lengths is ☐ cubic unit.

4. Find the product of the edge lengths of the $\frac{1}{2}$-unit cube and of the $\frac{1}{3}$-unit cube.

   $\frac{1}{2} \times \frac{1}{2} \times \frac{1}{2} = $ ☐

   $\frac{1}{3} \times \frac{1}{3} \times \frac{1}{3} = $ ☐

5. What do you observe about the volume of cubes with fractional edge lengths from **Question 3** and the product of the edge lengths from **Question 4**?

The results of **Class Activity 1** show that there are 8 cubes with $\frac{1}{2}$-unit side lengths in 1 cubic unit, hence the volume of one $\frac{1}{2}$-unit cube is $\frac{1}{8}$ cubic unit. And there are 27 cubes with $\frac{1}{3}$-unit side lengths in 1 cubic unit, hence the volume of one $\frac{1}{3}$-unit cube is $\frac{1}{27}$ cubic unit. We also see that the volume of cubes with fractional edge lengths has the same value as the product of the edge lengths.

Notice that both of these results agree with $V = lwh$ which was illustrated earlier for whole numbers.

Student Textbook page 157

4. **Extension**

Have students study Example 1 and do Try It! 1.

Discuss REMARKS.

Notes:
Example 1:

- Help students see that two $\frac{1}{3}$-m cubes will fit along each edge of the $\frac{2}{3}$-m cube, so it will take 2 × 2 × 2 = 8 cubes with $\frac{1}{3}$-m edges to fill the $\frac{2}{3}$-m cube. Each $\frac{1}{3}$-m cube has a volume of $\frac{1}{27}$ m³, so the volume of the $\frac{2}{3}$-m cube is $8 \times \frac{1}{27}$ m³ = $\frac{8}{27}$ m³.

- For Try It! 1, it takes 5 × 5 × 5 = 125 $\frac{1}{5}$-cm cubes to fill a 1-cm cube, thus the volume of a $\frac{1}{5}$-cm cube is $\frac{1}{125}$ cm³. Three $\frac{1}{5}$-cm cubes will fit along each edge of the $\frac{3}{5}$-cm cube so it will take 3 × 3 × 3 = 27 $\frac{1}{5}$-cm cubes to fill it. So, the volume of the $\frac{3}{5}$-cm cube is $27 \times \frac{1}{125}$ cm³ = $\frac{27}{125}$ cm³.

Try It! 1 Answer

$\frac{3}{5}$ cm × $\frac{3}{5}$ cm × $\frac{3}{5}$ cm = $\frac{27}{125}$ cm³

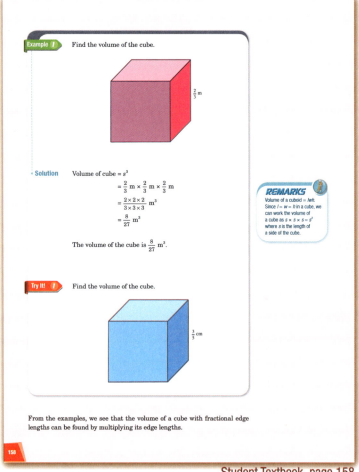

Student Textbook page 158

5. **Conclusion**

Summarize the main points of the lesson.
- A cuboid is a solid that is box-shaped, and is also called a right rectangular prism. To find the volume of a cuboid, we multiply length × width × height.
- Since the length, width, and height of a cube are the same, we only need to know the length of one edge. To find the volume of a cube, we can multiply length × width × height or edge × edge × edge. These formulas work even when the edges of the cube are fractional lengths.

200

Lesson 2

Objective: Find the volume of right rectangular prisms with fractional edge lengths.

1. Introduction

Tell students that in the previous lesson, we found that we can use the formula length × width × height to find the volume of a cube with fractional edge lengths. Today, we want to see if this formula will work for any right rectangular prism with fractional edge lengths.

2. Development

Pose the problem at the top of page 159 (cover the answer).

Look at the prism.

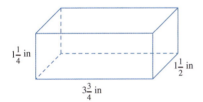

How many $\frac{1}{4}$-in cubes does it take to fill the prism?

Ask, "How many $\frac{1}{4}$-in cubes will fit along the length?" ($3\frac{3}{4} \div \frac{1}{4} = 15$)

Have students solve the problem on their own, discuss their answers with a partner or group, and share their solutions and methods with the class.

Read and discuss page 159. Discuss what the girl is saying.

Student Textbook page 159

Notes:

- Dividing each dimension by $\frac{1}{4}$ will give you the number of cubes that will fit across each.

Length → $3\frac{3}{4} \div \frac{1}{4} = 15$

Width → $1\frac{1}{2} \div \frac{1}{4} = 6$

Height → $1\frac{1}{4} \div \frac{1}{4} = 5$

- $4 \times 4 \times 4 = 64$ $\frac{1}{4}$-in cubes would fit in a 1-in cube. Thus, the volume of a $\frac{1}{4}$-in cube is $\frac{1}{64}$ in³. It will take $15 \times 6 \times 5 = 450$ $\frac{1}{4}$-in cubes to fill the cuboid so the volume of the cuboid is $450 \times \frac{1}{64}$ in³ $= \frac{450}{64}$ in³ $= 7\frac{1}{32}$ in³.

Use the top of page 160 to summarize. Discuss what the boy is saying.

3. Application

Have students study Examples 2 – 3 and do Try It! 2 – 3.

Notes:

Example 2:

- Instead of writing the unit of length (inch) after each number, we can use parentheses and just write the cubic unit (in³). For example, $\left(4 \times \frac{15}{4} \times \frac{13}{2}\right)$ in³.

- Generally, when you multiply fractions, it is best to convert the mixed numbers to improper fractions to do the calculation. In the final step, students should change the improper fraction to a mixed number in its simplest form as we do not normally express volume with improper fractions.

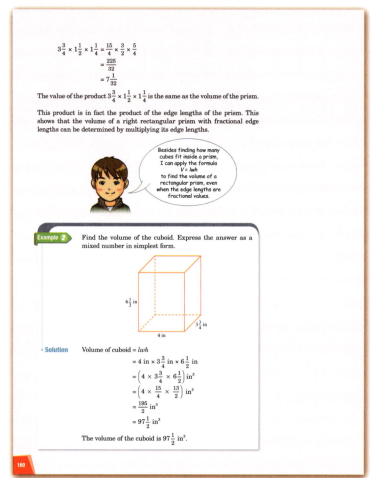

Student Textbook page 160

Try It! 2 Answer

$6 \text{ cm} \times 2\frac{1}{2} \text{ cm} \times 2\frac{1}{2} \text{ cm} = \left(6 \times \frac{5}{2} \times \frac{5}{2}\right) \text{ cm}^3$

$= \frac{75}{2} \text{ cm}^3$

$= 37\frac{1}{2} \text{ cm}^3$

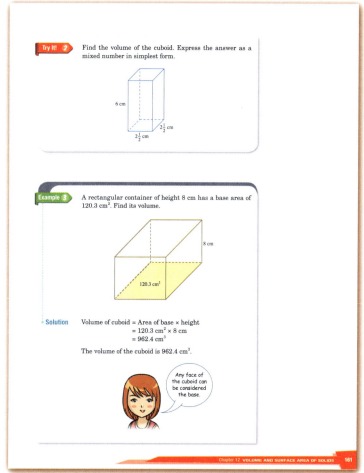

Example 3:
- Here, we are applying the formula Area of base × height to find the volume of a cuboid. Remind students that the area of the base = length × width, thus length × width = 120.3 cm².
- Point out that we use square units for the area of the base and cubic units for volume. A square unit × a linear unit = a cubic unit. E.g., for a 1-cm cube:

$1 \text{ cm} \times 1 \text{ cm} \times 1 \text{ cm} = (1 \text{ cm} \times 1 \text{ cm}) \times 1 \text{ cm}$
$= (1 \times 1) \text{ cm}^2 \times 1 \text{ cm}$
$= 1 \text{ cm}^3$

Notes:

- In Try It! 3, we are given the front face. Any face of a cuboid can be considered the base, so we can also use Area of base × height to find the volume. Thus,

  Volume of cuboid =
      Area of front (or back) face × width

  Volume of cuboid =
      Area of side face × length

  Volume of cuboid =
      Area of top (or bottom) face × height
      (this is the same as Area of base × height)

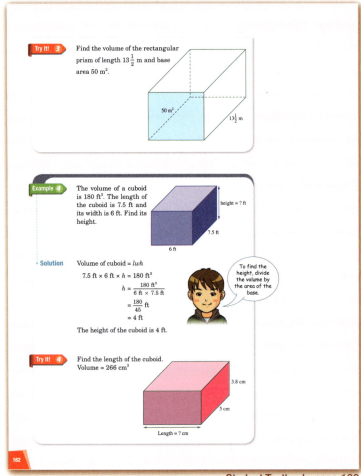

Student Textbook page 162

Try It! 3   Answer

$50 \text{ m}^2 \times 13\frac{1}{2} \text{ m} = 675 \text{ m}^3$

4. Extension

Here we are given the volume and two of the three dimensions. Since Volume = Area of base × height, height = $\frac{\text{Volume}}{\text{Area of base}}$. Thus, the following relationships hold true:

$$\text{Height} = \frac{\text{Volume}}{\text{Area of base}}$$

$$\text{Length} = \frac{\text{Volume}}{\text{width} \times \text{height}}$$

$$\text{Width} = \frac{\text{Volume}}{\text{length} \times \text{height}}$$

Try It! 4 **Answer**

$$5 \text{ cm} \times 3.8 \text{ cm} \times l = 266 \text{ cm}^3$$

$$l = \frac{266}{19} \text{ cm}$$

$$= 14 \text{ cm}$$

5. Conclusion

Summarize the main points of the lesson.
- We can find the volume of a cuboid with fractional edge lengths using the formula length × width × height.
- Since length × width = the area of the base, we can also use the formula Area of base × height.
- We can divide the volume of the cuboid by the area of one face (or by the product of two known dimensions) to find an unknown dimension.

Lesson 3

Objectives:
- Find the volume of compound solid figures composed of cuboids.
- Find the volume of liquids in a container.

1. Introduction

Show students 2 boxes (or other cuboids) placed or taped together (similar to Example 5).

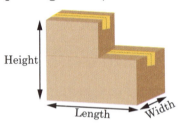

Ask, "Can we use the formula length × width × height to find the volume of this shape?"

Note: Help students see that $l \times w \times h$ would give the volume of a large box with those dimensions, but it will not work for this shape, because it is composed of two boxes.

2. Development

Have students solve Example 5 (cover the answer).

Notes:
- Remind students of the methods they used to find the areas of complex plane figures in Chapter 11; e.g., cutting it into smaller shapes and adding the areas, or creating a larger shape and subtracting out the extra area.
- If they find one way to solve the problem, ask them to think of another way.

Have students share and discuss their solutions and methods.

Discuss the solution given on page 163 and DISCUSS at the bottom right of the page.

Notes:
Example 5
- Additional methods:

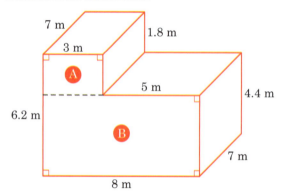

Method 2

Cut the two cuboids horizontally.

Volume of cuboid A =
$\quad$ 7 m × 3 m × 1.8 m = 37.8 m³
Volume of cuboid B =
$\quad$ 8 m × 4.4 m × 7 m = 246.4 m³
Volume of solid =
$\quad$ 246.4 m³ + 37.8 m³ = 284.2 m³

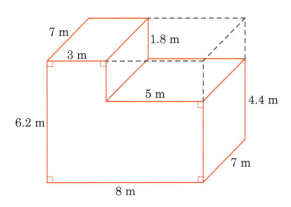

Method 3

Create a cuboid and subtract out the additional volume.

Volume of large cuboid =
 8 m × 7 m × 6.2 m = 347.2 m³

Volume of small cuboid =
 5 m × 7 m × 1.8 m = 63 m³

Volume of solid =
 347.2 m³ − 63 m³ = 284.2 m³

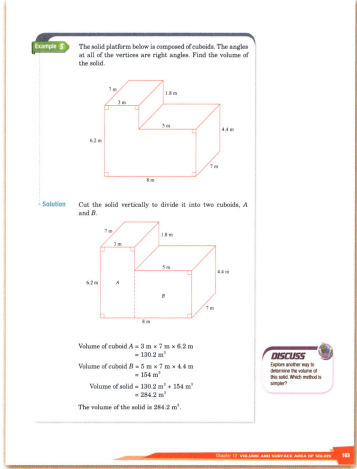

Student Textbook page 163

Have students do Try It! 5 and discuss the solution.

Try It! 5   Answer

Volume of cuboid A = 3 ft × 6 ft × $2\frac{1}{6}$ ft = 39 ft³

Volume of cuboid B = 10 ft × 3 ft × $3\frac{3}{4}$ ft = $112\frac{1}{2}$ ft³

Volume of solid = 39 ft³ + $112\frac{1}{2}$ ft³ = $151\frac{1}{2}$ ft³

- For Try It! 5, we can also cut the shape vertically into three cuboids. Point out that although this will work, if we cut the shape horizontally it is easier because there will only be two cuboids to find the volume of.

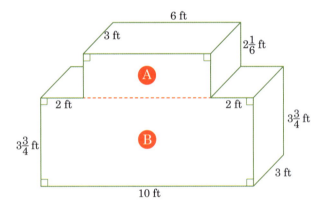

- Students could also imagine a large cuboid by adding two $(2 \times 3 \times 2\frac{1}{6})$ ft cuboids to the top right and top left sides, find the volume of the large cuboid, and then subtract the volume of the two smaller cuboids.

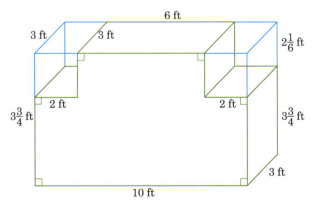

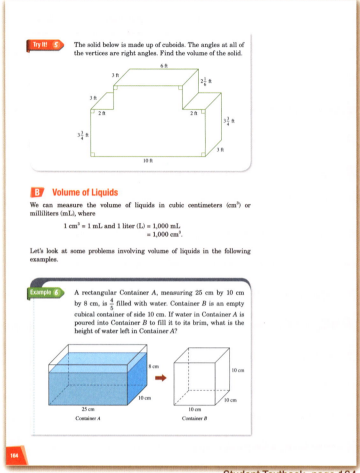

Student Textbook page 164

Volume of large cuboid =

$10 \times 3 \times \left(3\frac{3}{4} + 2\frac{1}{6}\right) = 177\frac{1}{2}$ ft³

Volume of two small cuboids =

$2 \times 3 \times 2\frac{1}{6} \times 2 = 26$ ft³

Volume of solid figure = $177\frac{1}{2} - 26 = 151\frac{1}{2}$ ft³

★ **Workbook: Page 93**

3. Application

Discuss the conversions of cm³ and mL on the middle of page 164.

Have students study Example 6, do Try It! 6, and discuss the solution.

Notes:

Example 6:
- These problems relate the volume of cuboids to real life examples involving capacity and liquid volume. Students learned these conversions in Grade 5.
- It may be helpful to show a concrete example by pouring water from a larger cube-shaped container into a smaller one.

Remind students that since length × width × height = Volume, height = Volume ÷ (length × width) or Volume ÷ Area of the base.

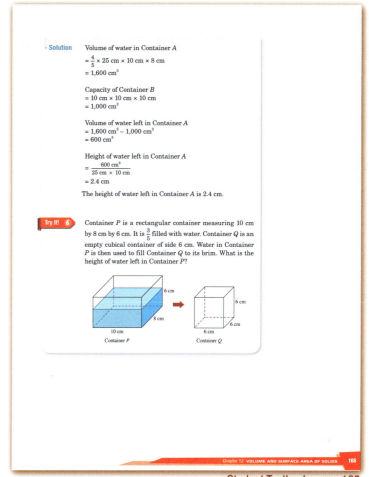

Student Textbook page 165

Try It! 6 Answer

Volume of water in Container P =

$$\frac{3}{5} \times 10 \text{ cm} \times 8 \text{ cm} \times 6 \text{ cm} = 288 \text{ cm}^3$$

Capacity of Container Q =

$$6 \text{ cm} \times 6 \text{ cm} \times 6 \text{ cm} = 216 \text{ cm}^3$$

Volume of water left in Container P =

$$288 \text{ cm}^3 - 216 \text{ cm}^3 = 72 \text{ cm}^3$$

Height of water left in Container P =

$$\frac{72 \text{ cm}^3}{10 \text{ cm} \times 8 \text{ cm}} = \frac{9}{10} \text{ cm, or } 0.9 \text{ cm}$$

4. **Extension**

Discuss REMARKS and have students study Example 7.

Have students do Try It! 7.

Notes:

Example 7:
- Since it takes 15 min to fill the tank at the given rate of flow, we can find the capacity of the tank by multiplying the rate of flow by 15.
- Here, students need to convert from liters to cubic centimeters. It may be helpful to convert to milliliters first; e.g., 150 L = (150 × 1,000) mL = 150,000 mL = 150,000 cm^3.
- For Try It! 5, we need to find the volume of water in the tank and then divide by the rate of flow to find how long it will take to empty the tank. To do that, we need to find the volume of the water in the tank in cm^3 first and then convert it to liters.

Try It! 7 Answer

Volume of water in tank =

$\frac{2}{3}$ × 45 cm × 30 cm × 35 cm = 31,500 cm^3

31,500 cm^3 = $\frac{31,500}{1,000}$ L = 31.5 L

$\frac{31.5 \text{ liters}}{5 \text{ liters per min}}$ = 6.3 minutes

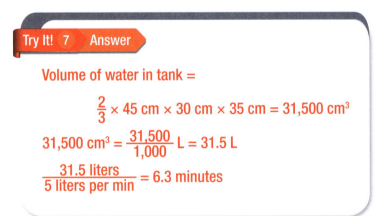

5. **Conclusion**

Summarize the important points of the lesson.
- We can find the volume of complex solid figures by cutting the solid into smaller solids and adding the volumes, or adding additional volume to create a larger solid and

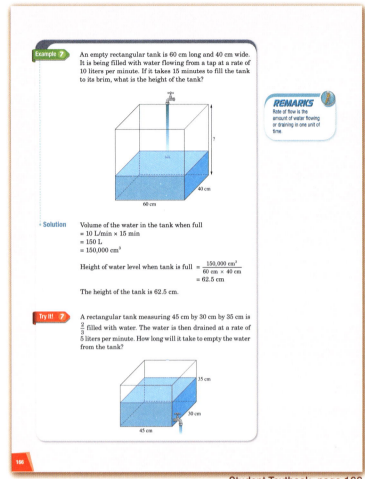

Student Textbook page 166

subtracting out the extra volume. These are the same methods we used to find the area of complex plane figures.
- We can convert between cubic centimeters and liters to solve real-life problems involving liquid volume.
- We can consider a rectangular tank with water in it to be two cuboids stacked on each other and apply the idea that the total volume of the tank is the sum of the volume with water and the volume without water.

★ **Workbook: Page 101**

Lesson 4

Objective: Consolidate and extend the material covered thus far.

Have students work together with a partner or in groups. Students should try to solve the problems by themselves first, then compare solutions with their partner or group. If they are confused, they can discuss together.

Observe students carefully as they work on the problems. Give help as needed individually or in small groups.

Note: Due to the complexity of the problems, students may use a calculator at the teacher's discretion, especially for problems 12 – 17.

BASIC PRACTICE

1. (a) 76.5 cm³   (b) $32\frac{1}{2}$ in³

2. (a) 42.875 cm³   (b) $15\frac{5}{8}$ ft³

   (c) $\frac{27}{64}$ in³

3. $\frac{5}{6} \times$ 22 cm × 6 cm × 10 cm = 1,100 cm³

FURTHER PRACTICE

4. (a) 36 in² × $5\frac{3}{4}$ in = 207 in³

   (b) 23.5 m² × 9 m = 211.5 m³

   (c) 12.5 cm² × 4.6 cm = 57.5 cm³

5. (a) $\frac{420 \text{ m}^3}{12 \text{ m} \times 7 \text{ m}}$ = 5 m

   (b) $\frac{217 \text{ cm}^3}{5 \text{ cm} \times 2.8 \text{ cm}}$ = 15.5 cm

   (c) $\frac{180 \text{ in}^3}{10 \text{ in} \times 4.5 \text{ in}}$ = 4 in

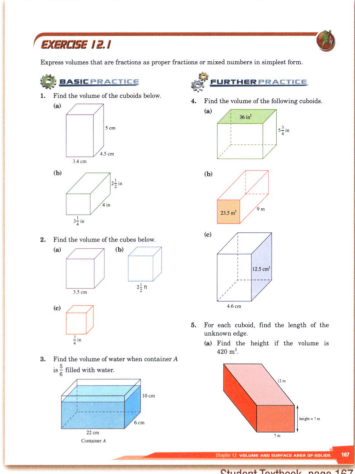

Student Textbook page 167

6. (a) Volume of cuboid A =
      (3 cm + 2 cm + 3 cm) × 4 cm × 1.5 cm = 48 cm³
   Volume of cuboid B =
      2 cm × 5 cm × 1.5 cm = 15 cm³
   Volume of solid =
      48 cm³ + 15 cm³ = 63 cm³

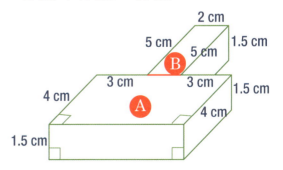

(b) Volume of cuboid A = 4 in × $2\frac{1}{2}$ in × 2 in = 20 in³

Volume of cuboid B = 4 in × $2\frac{1}{2}$ in × 2 in = 20 in³

Volume of cuboid C =

10 in × ($6\frac{1}{4}$ in – 2 in) × $2\frac{1}{2}$ = $106\frac{1}{4}$ in³

Volume of solid =

20 in³ + 20 in³ + $106\frac{1}{4}$ in³ = $146\frac{1}{4}$ in³

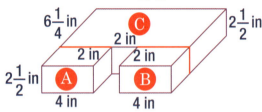

- Students could also convert the mixed numbers to decimals to make the calculations easier and get 146.25 in³.
- The answer for 6 (b) in the back of the book gives the answer $148\frac{1}{4}$ in³, which is wrong.

7. 7.5 L = 7,500 mL = 7,500 cm³

$$\frac{7{,}500 \text{ cm}^3}{50 \text{ cm} \times 30 \text{ cm}} = 5 \text{ cm}$$

8.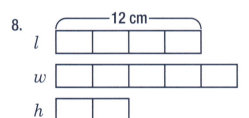

Length (4 units) ⟶ 12 cm

1 unit ⟶ 12 cm ÷ 4 = 3 cm

Width (5 units) ⟶ 3 cm × 5 = 15 cm

Height (2 units) ⟶ 3 cm × 2 = 6 cm

Volume ⟶ 12 cm × 15 cm × 6 cm = 1,080 cm³

9. L ▢▢
   W ▢
   H ▢▢▢
   └─15 cm─┘

   Height (3 units) ⟶ 15 cm
   Width (1 unit) ⟶ 15 cm ÷ 3 = 5 cm
   Length (2 units) ⟶ 5 cm × 2 = 10 cm
   Volume ⟶ 10 cm × 5 cm × 15 cm = 750 cm³

 MATH@WORK

10. Volume of Box − Volume of packing foam = space left

    $\left(3\frac{1}{2} \text{ ft} \times 2\frac{3}{4} \text{ ft} \times 4 \text{ ft}\right) - 10 \text{ ft}^3 = 28\frac{1}{2} \text{ ft}^3$

11. (a)

	Length (ft)	Width (ft)	Height (ft)	Volume (ft³)
Box A	2	$1\frac{1}{2}$	1	3
Box B	2	$1\frac{1}{3}$	$1\frac{1}{3}$	$3\frac{5}{9}$
Box C	$1\frac{2}{3}$	$1\frac{3}{4}$	$1\frac{1}{2}$	$4\frac{3}{8}$

(b) $4\frac{3}{8}$ ft³ − 3 ft³ = $1\frac{3}{8}$ ft³

(c) Box A and Box B are too small because 3.6 is greater than 3 and $3\frac{5}{9}$.

$4\frac{3}{8} > 3.6$, so they should choose Box C.

12. Method 1:
    Volume of water in container at first =
    30 cm × 25 cm × 4.5 cm = 3,375 cm³
    Water poured in = 3 L = 3,000 mL = 3,000 cm³
    Volume of water in container after =
    3,375 cm³ + 3,000 cm³ = 6,375 cm³
    Height = $\frac{6{,}375 \text{ cm}^3}{30 \text{ cm} \times 25 \text{ cm}}$ = 8.5 cm

    Method 2:
    3,000 ÷ (30 × 25) = 4 cm
    4.5 + 4 = 8.5 cm

13. Method 1:
    Volume of water in container at first =
    32 cm × 28 cm × 19 cm = 17,024 cm³
    Water removed = 5.6 L = 5,600 mL = 5,600 cm³
    Volume of water in container after =
    17,024 cm³ − 5,600 cm³ = 11,424 cm³
    Height = $\frac{11{,}424 \text{ cm}^3}{32 \text{ cm} \times 28 \text{ cm}}$ = 12.75 cm

    Method 2:
    5,600 ÷ (32 × 28) = 6.25 cm
    19 − 6.25 = 12.75 cm

14. Method 1:
    Capacity of tank =
    40 cm × 35 cm × 16 cm = 22,400 cm³
    Volume of water in tank =
    8.82 L = 8,820 mL = 8,820 cm³
    Volume of empty space in tank =
    22,400 cm³ − 8,820 cm3 = 13,580 cm³
    Distance from water level to top of tank =
    $\frac{13{,}580 \text{ cm}^3}{40 \text{ cm} \times 35 \text{ cm}}$ = 9.7 cm

    Method 2:
    8,820 ÷ (40 × 35) = 6.3 cm
    16 − 6.3 = 9.7 cm

15. Method 1:
    Volume of water in container =
    28 cm × 25 cm × 14 cm = 9,800 cm³
    Additional water needed to fill the tank =
    4 L 60 mL = 4,000 cm³ + 60 cm³ = 4,060 cm³
    Capacity of tank = 9,800 cm³ + 4,060 cm³ = 13,860 cm³
    Height = $\frac{13{,}860 \text{ cm}^3}{28 \text{ cm} \times 25 \text{ cm}}$ = 19.8 cm

    Method 2:
    4,060 ÷ (28 × 25) = 5.8 cm
    14 + 5.8 = 19.8 cm

16. Method 1:

    Capacity of tank = 54 L ÷ $\frac{3}{4}$ = 72 L = 72,000 cm³

    height = $\frac{72,000 \text{ cm}^3}{60 \text{ cm} \times 32 \text{ cm}}$ = $37\frac{1}{2}$ cm

    Method 2:

    $\frac{3}{4}$ of the height of the tank =
    54,000 ÷ (60 × 32) = $28\frac{1}{8}$ cm

    $\frac{1}{4}$ of the height of the tank =
    $28\frac{1}{8}$ ÷ 3 = $9\frac{3}{8}$ cm

    The height of the tank = $9\frac{3}{8}$ × 4 = $37\frac{1}{2}$ cm

17. Capacity of tank =
    88 cm × 60 cm × 40 cm = 211,200 cm³ = 211.2 L

    211.2 ÷ 12 = 17.6

    It will take 17.6 minutes.

BRAIN WORKS

12. A rectangular container is 30 cm long and 25 cm wide. The height of the water level in the container is 4.5 cm. Three liters of water are then poured into the container. What is the height of the water level now?

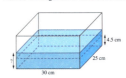

13. A rectangular container is 32 cm long and 28 cm wide. The height of the water level in the container is 19 cm. After 5.6 L of water are removed from the container, what is the height of the water level?

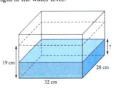

14. A rectangular tank is 40 cm long, 35 cm wide, and 16 cm high. It is filled with 8.82 L of water. How far is the water level, measured from the top of the tank?

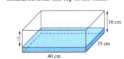

15. A rectangular container is 28 cm long, and 25 cm wide. The height of the water level in the container is 14 cm. To fill the container to its brim, 4 L 60 mL of water are poured into it. What is the height of the container?

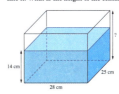

16. A rectangular tank is 60 cm long and 32 cm wide. It contains 54 liters of water when it is $\frac{3}{4}$ full. Find the height of the tank.

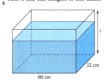

17. An empty rectangular tank measuring 88 cm by 60 cm by 40 cm. It is being filled with water flowing from a tap at a rate of 12 liters per minute. How long will it take to fill the tank to its brim?

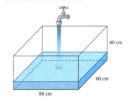

Student Textbook page 169

Lesson 5

Objectives:
- Identify various types of prisms.
- Find the surface area of right rectangular prisms.

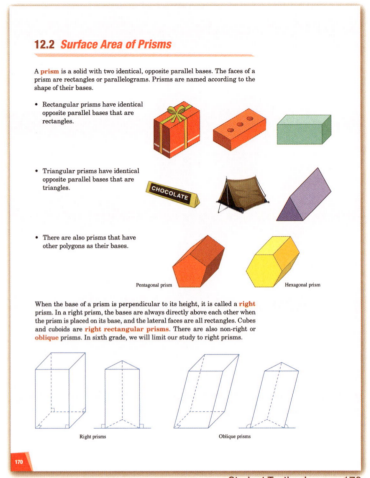

Student Textbook page 170

1. Introduction

Give groups of students some examples of different kinds of prisms to examine.
- These can be models of prisms or real-life examples similar to those on page 170.

Read page 170 and discuss the similarities and differences of the prisms.

Notes:
- The bases of prisms do not have to be convex shapes.
- On right prisms, the height is perpendicular to both bases (i.e. the lateral faces and the bases meet at right angles). The example of oblique prisms is given so students can see that not all prisms have heights that are perpendicular to the bases. However, the study of prisms in **Dimensions Math® 6** will be limited to right rectangular and right triangular prisms.

2. Development

Read the top of page 171. Ask students to give real life examples of surface area; e.g., painting a house, wrapping a present, building a box.

Give students graph paper and rulers, and have them draw the net of the cuboid on page 171 and label the faces.

Have students cut out the net and fold it at the vertices to form a cuboid. (Do not tape the vertices.)

Ask students which edges have the same length and why.

Note: Help students see that the sides of the faces that share a common edge are the same length; e.g., the left side of face *A* is the same length as the top side of face *D*, because they share the same edge.

Have students work with a partner or in groups to complete Class Activity 2.

Answers for Class Activity 2

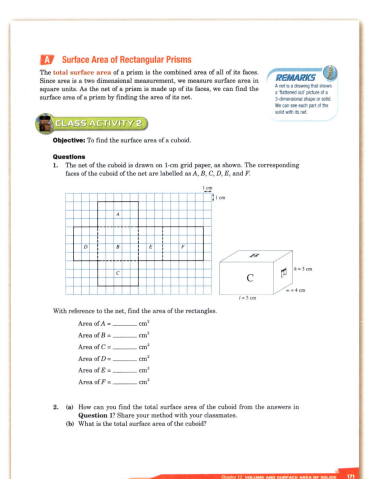

Student Textbook page 171

1. 15
   20
   15
   12
   12
   20

2. (a) Add the areas of all the faces.

   (b) 15 cm² + 20 cm² + 15 cm² + 12 cm² + 12 cm² + 20 cm² = 94 cm²

3. (a) Bottom and top

   (b) Left and right

   (c) Front and back

4. (a) The top (B) and bottom (F). 40 cm²

   (b) 30 cm², front, back

   (c) 24 cm², right, left

5. (5 cm × 4 cm × 2) + (5 cm × 3 cm × 2) + (3 cm × 4 cm × 2) = 40 cm² + 30 cm² + 24 cm²

   = 94 cm²

6. (a) They are the same.

   (b) You only have to find the areas of 3 of the faces. Since the opposite faces have the same area, we can multiply each face by 2.

Have students share and explain their answers to Class Activity 2.

Use the bottom of page 172 to summarize the activity.

Notes:

- Help students see that they cannot just find the areas of any three faces (e.g., top, bottom, and right) but the areas of the unique faces (e.g., top, front and left, or bottom, back, and right).

- Instead of finding the area of each unique face and multiplying by 2, they can add the areas of the three unique faces and multiply the total area by 2.

---

3. Name the three pairs of identical faces of a cuboid. One pair is already given.
   (a) bottom and top faces
   (b) _____ and _____ faces
   (c) _____ and _____ faces

4. (a) Which two faces of the cuboid represent the area found by multiplying length by width? How many cm² is $l \times w \times 2$?
   (b) $l \times h \times 2 =$ _____ cm² gives the combined area of the _____ and _____ faces.
   (c) $w \times h \times 2 =$ _____ cm² gives the combined area of the _____ and _____ faces.

5. Using your answers from **Question 4**, find the combined area of all faces of the cuboid.

6. (a) What do you notice about the total surface area of the cuboid from **Question 2** and $(l \times w \times 2) + (l \times h \times 2) + (w \times h \times 2)$?
   (b) What conclusion can you draw?

We see from **Class Activity 2** that to find the total surface area of a cuboid, we need to find the sum of the areas of all the faces. Since the opposite faces of a cuboid have the same area, we only need to know the areas of three of the faces.

Total surface area of a cuboid
= (Area of base × 2) + (Area of front or back face × 2) + (Area of side face × 2)
= ($l \times w \times 2$) + ($l \times h \times 2$) + ($w \times h \times 2$)
= ($l \times w + l \times h + w \times h$) × 2
= ($lw + lh + wh$) × 2

Instead of multiplying each area by 2 separately, we can also add the three areas and multiply their sum by 2.

Total surface area of a cuboid
= (Area of base + Area of front or back face + Area of side face) × 2
= ($l \times w + l \times h + w \times h$) × 2
= ($lw + lh + wh$) × 2

REMARKS
This formula gives us the sum of the areas of all the faces of the cuboid.

Total surface area of a cuboid = ($lw + lh + wh$) × 2

Student Textbook page 172

3. Application

Have students study Example 8 and do Try It! 8, and then discuss the solution.

Note: Method 2 is really a simplified version of Method 1. Help students see how the methods are related. (e.g., $5 \times 6 + 5 \times 4 + 6 \times 4$ is the area of base + area of front face + area of side face).

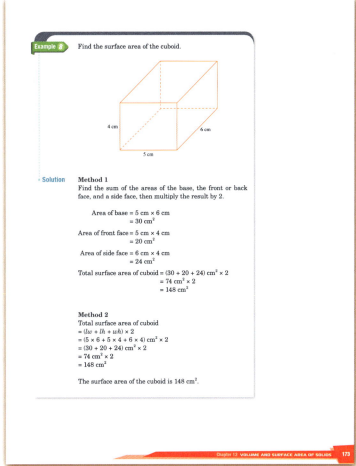

Student Textbook page 173

Try It! 8   Answers

(a) Area of base = 6 ft × 4 ft = 24 ft²
Area of front face = 6 ft × 9 ft = 54 ft²
Area of side face = 4 ft × 9 ft = 36 ft²
Total area of cuboid =
(24 + 54 + 36) ft² × 2 = 228 ft²

(b) Total surface area of cuboid
$= (lw + lh + wh) \times 2$
$= (3 \times 10 + 3 \times 2\frac{1}{2} + 10 \times 2\frac{1}{2}) \text{ in}^2 \times 2$
$= (30 + 7\frac{1}{2} + 25) \text{ in}^2 \times 2$
$= 62\frac{1}{2} \text{ in}^2 \times 2$
$= 125 \text{ in}^2$

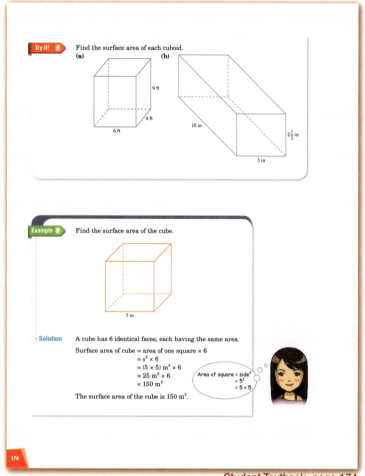

Student Textbook page 174

4. Extension

Have students study Example 9 and do Try It! 9.

Notes:
- The surface area of a cube is a special case because we only need to know the area of one face. Since all the faces are congruent, we multiply the area of one face by 6.
- Help students see that since a cube is also a cuboid, we could also use the formula for the surface area of a cuboid; (5 × 5 + 5 × 5 + 5 × 5) × 2 but this method is more difficult.

Lesson 6

Try It! 9 Answers

(a) (20 × 20) mm² × 6 = 2,400 mm²

Students could convert mm to cm to find the surface area.
20 mm = 2 cm
(2 × 2) × 6 = 24 cm²

(b) $\left(\frac{3}{4} \times \frac{3}{4}\right)$ m² × 6 = $3\frac{3}{8}$ m²

5. Conclusion

Summarize the main points of the lesson.
- The surface area of a solid is the combined area of all of its faces.
- For cuboids, 3 faces are unique and 3 faces are duplicates so we only need to find the areas of 3 faces and multiply by 2.
- To find the surface area of a cube, we only need to find the area of one face and multiply by 6.

★ **Workbook: Page 110**

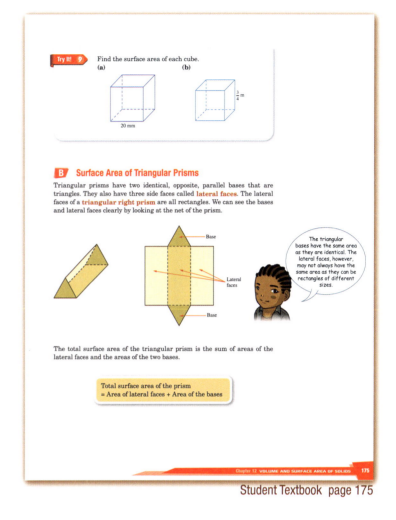

Student Textbook page 175

Objective: Find the surface area of triangular prisms.

1. Introduction

Show students a triangular prism.

Ask, "How is the surface area of the triangular prism similar or different from the surface area of a rectangular prism?"

Read and discuss page 175.

Note: Help students see that the formula given for finding the surface area of a triangular prism also holds for a rectangular prism. This formula will work for any prism, regardless of the shape of the base.

2. Development

Give students graph paper and rulers and have them draw the net of the cuboid on page 176 and label the faces.

Have students cut out the net and fold it at the vertices to form a triangular prism. (Do not tape the vertices.)

Have students work with a partner or in groups to complete Class Activity 3.

Answers for Class Activity 3

1. Area of $A = \frac{1}{2} \times 3$ cm $\times 4$ cm $= 6$ cm^2

   Area of $B = \frac{1}{2} \times 3$ cm $\times 4$ cm $= 6$ cm^2
   Area of $C = 5$ cm $\times 7$ cm $= 35$ cm^2
   Area of $D = 4$ cm $\times 7$ cm $= 28$ cm^2
   Area of $E = 3$ cm $\times 7$ cm $= 21$ cm^2
   Notes:
   - Tell students to write out the calculations for how they found the area of each face.
   - Ask, "Why do the calculations for finding the areas of C, D, and E all have "times 7"? (Because the horizontal side of all of the rectangular faces is the same as the height of the prism; 7 cm.)
   Ask: "What sides of triangle A are the same length as QR, QS and ST? ($QR = PQ$, $QS = QS$, $RS = ST$)

2. Add the areas of all the faces.
   $(6 + 6 + 35 + 28 + 21)$ cm^2 $= 96$ cm^2

3. 6 cm$^2 + 6$ cm$^2 = 6$ cm$^2 \times 2 = 12$ cm^2

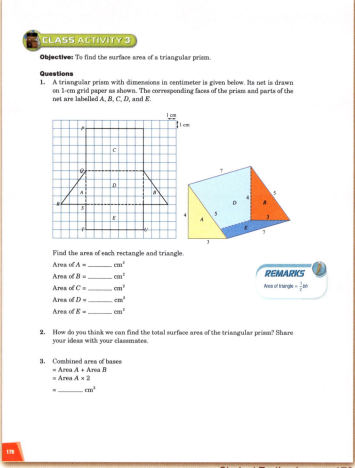

Student Textbook page 176

4. (a) Method 1 | Method 2
   D, E | 3, 4, 5
   35, 28, 21 | 12
   84 | PT
   | 12, 84, PT

If students are having difficulty finding the length of QR, point out that side QR on the triangle forms a common edge with PQ on the rectangle, so they are the same length (5 cm).

(b) A common edge is formed where the triangular base meets the rectangular lateral face. That means the length of the triangular edge and the width of its adjoining face are the same.

QR on triangle $A = PQ$ on rectangle C
QS on triangle $A = QS$ on rectangle D
RS on triangle $A = ST$ on rectangle E

Thus, multiplying the width of each rectangle by the height is the same as multiplying one side of the triangular base by the height.

$$PQ \times h + QS \times h + ST \times h = QR \times h + QS \times h + RS \times h$$
$$= (QR + QS + RS) \times h$$
$$= 3 \times 7 + 4 \times 7 + 5 \times 7$$
$$= (3 + 4 + 5) \times 7$$

5. 84, 12
   96

Have students share and explain their answers to Class Activity 3. Use the bottom of page 177 to summarize the activity.

Note: The formula Perimeter of base × height + base area × 2 will find the surface area of any prism, including rectangular prisms (See REMARKS).

4. Find the lateral surface area in the following ways.
   (a) Method 1
       Add the areas of the lateral faces of the prism.
       Lateral surface area
       = Area of C + Area of _____ + Area of _____
       = _____ cm² + _____ cm² + _____ cm²
       = _____ cm²

       Method 2
       Add the lengths of the three sides of face A to find the perimeter of the triangular base.
       Perimeter of base = _____ cm + _____ cm + _____ cm
       = _____ cm
       With reference to the net,
       Perimeter of the triangular base = length of line segment _____.
       Multiply the perimeter of the base by the height of the prism.
       _____ cm × 7 cm = _____ cm², that is _____ × TU.

   (b) Explain why in **Method 2**, "perimeter of base × height" gives the lateral surface area as calculated in **Method 1**.

5. Total surface area of the prism
   = Lateral surface area + Areas of the bases
   = _____ cm² + _____ cm²
   = _____ cm²

We can find the total surface area of a triangular prism, by adding the lateral surface area with the combined areas of the two bases.

We see from **Class Activity 3** that the sides of the triangular base and the height of the prism are the length and width of each rectangular lateral face. Thus, we can find the lateral surface area of a prism by multiplying the perimeter of the base by the height.

Since the two bases are identical triangles, we can find the combined areas of the bases by finding the area of one base and multiplying it by 2.

Total surface area of prism
= Perimeter of base × Height + Base area × 2

REMARKS
This formula applies to all prisms, not just to a triangular prism.

Student Textbook page 177

3. **Application**

Have students study Example 10 and do Try It! 10.

Notes:

Example 10:
- The base is an isosceles triangle. Since two sides of the triangle have the same length, the right and left face are identical rectangles. (See REMARKS.)
- Method 2 is really just a shorter version of Method 1. To find the lateral surface area using Method 1, we multiply the base of the triangle by height, the left side of the triangle by the height and the right side of the triangle by the height, and add the areas. Since the height is a common factor we can just multiply the sum of the sides of the triangle by the height.
- In Try It! 10 (a), students may be confused about which faces are the bases because the prism is sitting on a rectangular face. Remind them that the bases of a triangular prism are always the two triangular faces.
- In Try It! 10 (b), the prism is oriented differently. In this case, Method 2 is probably easiest because it may be difficult to see the dimensions of each lateral face.

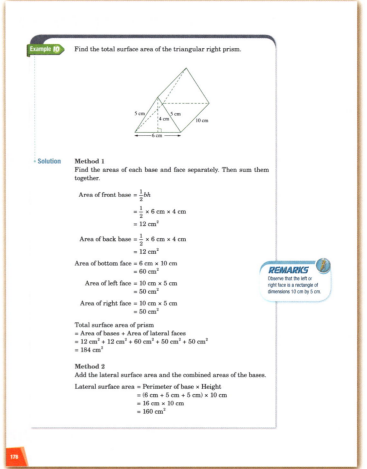

Student Textbook page 178

4. **Conclusion**

Summarize the main points of the lesson.
- We can find the surface area of any prism, regardless of the shape of the base, by adding the lateral surface area and the combined area of the two bases.
- To find the lateral surface area, we can multiply the length and width of each rectangular lateral face and then add the areas. An easy way to do this is to just multiply the perimeter of a base by the height.

Try It! 10 Answers

(a) Method 1

Area of front base = $\frac{1}{2}$ × 5 cm × 12 cm = 30 cm^2

Area of back base = 30 cm^2

Area of bottom face = 5 cm × $4\frac{1}{2}$ cm = $22\frac{1}{2}$ cm^2

Area of left face = 12 cm × $4\frac{1}{2}$ cm = 54 cm^2

Area of right face = 13 cm × $4\frac{1}{2}$ cm = $58\frac{1}{2}$ cm^2

Total surface area =
$(30 + 30 + 22\frac{1}{2} + 54 + 58\frac{1}{2})$ cm^2 = 195 cm^2

Method 2

Perimeter of base × height + base area × 2

$= (12\text{ cm} + 5\text{ cm} + 13\text{ cm}) \times 4\frac{1}{2} +$
$(\frac{1}{2} \times 12\text{ cm} \times 5\text{ cm}) \times 2$

$= (30 \times 4\frac{1}{2})$ cm^2 + (30×2) cm^2 =
135 cm^2 + 60 cm^2
= 195 cm^2

(b) Method 1

Area of front base = $\frac{1}{2}$ × 18 m × 12 m = 108 m^2

Area of back base = 108 m^2

Area of top face = 18 m × 40 m = 720 m^2

Area of left face = 40 m × 15 m = 600 m^2

Area of right face = 40 m × 15 m = 600 m^2

Total surface area =
(108 + 108 + 720 + 600 + 600) m^2 = 2,136 m^2

Method 2

Perimeter of base × height + base area × 2

= (18 m + 15 m + 15 m) × 40 +
$(\frac{1}{2} \times 18\text{ m} \times 12\text{ m}) \times 2$

= (48 × 40) m^2 + (108 × 2) m^2 =
1,920 m^2 + 216 m^2
= 2,136 m^2

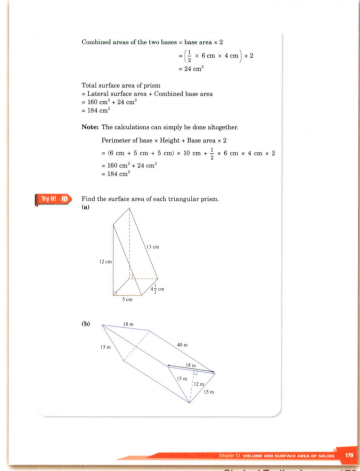

Student Textbook page 179

★ **Workbook: Page 119**

Lesson 7

Objectives:
- Convert between metric units of area and volume.
- Solve problems involving area, volume, and surface area that require measurement conversions.

1. Introduction

Make a square on the floor using 4 meter sticks.

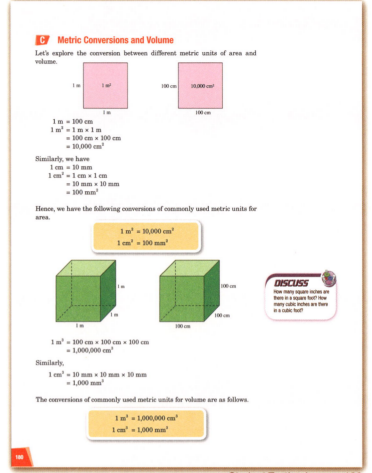

Student Textbook page 180

Ask, "What is the area of the 1-m square in square centimeters?"
- Each side of the 1-m square is 100 cm, so it's 100 cm × 100 cm = 10,000 cm².

Give groups or pairs of students a 1-cm grid paper. Ask, "What is the area of a 1-cm square in square millimeters?"
- Each side of the 1-cm square is 10 mm, so it's 10 mm × 10 mm = 100 mm².

Notes:
- To convert from a larger unit to a smaller unit, we multiply by the conversion factor. For example, 5 m = (5 × 100) cm = 500 cm.
- To convert from a smaller unit to a larger unit, we divide by the conversion factor. For example, 500 cm = (500 ÷ 100) m = 5 m.

This is the same as multiplying by a unit fraction in which the denominator is the conversion factor. For example:

$$500 \text{ cm} = \left(500 \times \frac{1}{100}\right) \text{ m} = 5 \text{ m}$$

- We can extend these ideas to area and volume by multiplying by the appropriate conversion factor. For example:

$$5 \text{ m}^2 = (5 \times 10{,}000) \text{ cm}^2 = 50{,}000 \text{ cm}^2$$

$$50{,}000 \text{ cm}^2 = \left(50{,}000 \times \frac{1}{10{,}000}\right) \text{ m}^2 = \frac{50{,}000}{10{,}000} \text{ m}^2 = 5 \text{ m}^2$$

$$5 \text{ m}^3 = (5 \times 1{,}000{,}000) \text{ cm}^3 = 5{,}000{,}000 \text{ cm}^3$$

$$5{,}000{,}000 \text{ cm}^3 = \left(5{,}000{,}000 \times \frac{1}{1{,}000{,}000}\right) \text{ m}^3 = \frac{5{,}000{,}000}{1{,}000{,}000} \text{ m}^3 = 5 \text{ m}^3$$

2. Development

Have several students help you to make a 1-m cube with 12 meter sticks.

Use the square meter you made on the floor and then place the other meter sticks vertically and horizontally to make all the edges of the cube. Ask some students to help by holding the meter sticks in place.

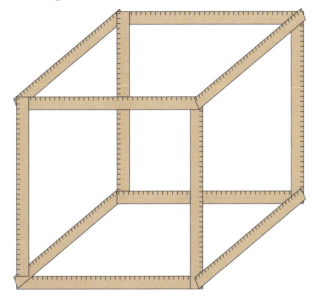

Give each student a 1-cm cube.

Ask, "How many 1-cm cubes will it take to fill a 1-m cube?"

Let students discuss this with a partner or in groups and share their ideas.

Possible responses:
- 100 1-cm cubes fit along each dimension (length, width, and height), so we can multiply $100 \times 100 \times 100 = 1{,}000{,}000$ 1-cm cubes.
- The area of the base is 100 cm × 100 cm = 10,000 cm². Area of base × height = 10,000 cm² × 100 cm = 1,000,000 cm³, so it takes 1,000,000 1-cm cubes.

Ask, "How many 1-mm cubes do you think would fit in a 1-cm cube?" Allow students to discuss this with a partner or in groups and share their ideas.
- The edges of a 1-cm cube are each 1 cm, which equals 10 mm. Thus, the volume is 10 mm × 10 mm × 10 mm = 1,000 mm³, so it's 1,000 1-mm cubes.

Read and discuss page 180. Ask students to talk about DISCUSS with partners or in groups.
- The dimensions of a 1-ft cube are 12 in by 12 in by 12 in, so there are 1,728 in³ in 1 ft³.

Notes:
- Again, to convert from a larger unit to a smaller unit, we multiply by the conversion factor; e.g., 5 m³ = (5 × 1,000,000) cm³ = 5,000,000 cm³.
- To convert from a smaller unit to a larger unit, we divide by the conversion factor; e.g., 5,000,000 cm³ = (5,000,000 ÷ 1,000,000) m³ = 5 m³.

3. Application

Have students study Examples 11 – 12 and do
Try It! 11 – 12.

Notes:

Example 11:

- For Type A, we are given the area in a
  smaller unit and we need to convert it to a
  larger unit. To convert from a smaller unit
  to a larger unit, we divide by the conversion
  factor because there will be fewer of the
  larger unit.

$$1{,}950 \text{ mm}^2 = \left(1{,}950 \times \tfrac{1}{100}\right) \text{cm}^2 = 19.5 \text{ cm}^2$$

Multiplying by $\tfrac{1}{100}$ is the same as dividing by

100, so students can divide 1,950 mm^3 by 100

instead of multiplying by $\tfrac{1}{100}$ if it is easier. To

divide by 100 $\left(\text{or multiply by } \tfrac{1}{100}\right)$, just move

the decimal point two places to the left.

- For Type B, we are given the area in a larger
  unit and we need to convert it to a smaller
  unit. To convert from a larger unit to a
  smaller unit, we multiply by the conversion
  factor because there will be more of the
  smaller unit.

$$0.5 \text{ cm}^3 = (0.5 \times 10{,}000) \text{ m}^3 = 5{,}000 \text{ cm}^3$$

To multiply by 10,000, we move the decimal
point 4 places to the right.

Try It! 11 Answers

(a) $1 \text{ mm}^2 = \frac{1}{100} \text{ cm}^2$

$1{,}245 \text{ mm}^2 = \left(\frac{1}{100} \times 1{,}245\right) \text{ cm}^2 = 12.45 \text{ cm}^2$

$10{,}000 \text{ cm}^2 = 1 \text{ m}^2$

$1 \text{ cm}^2 = \frac{1}{10{,}000} \text{ m}^2$

$12.45 \text{ cm}^2 = \left(\frac{1}{10{,}000} \times 12.45\right) \text{ m}^2 = 0.001245 \text{ m}^2$

(b) $10{,}000 \text{ cm}^2 = 1 \text{ m}^2$

$1 \text{ cm}^2 = \frac{1}{10{,}000} \text{ m}^2$

$368 \text{ cm}^2 = 0.0368 \text{ m}^2$

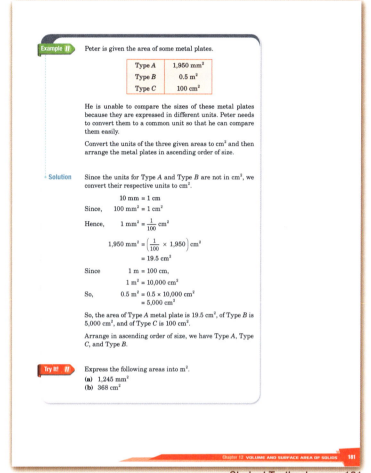

Student Textbook page 181

Notes:

Example 12:

- In these problems, students have to find both the volume and surface area in two different units.

- In Example 12 (b), to find in square centimeters, instead of converting to cm and applying the formula again, we could just convert the volume we found in m³ to cm³.
0.00054 m³ = (0.00054 × 1,000,000) cm³ = 540 cm³

- Regarding the blue text ("Alternatively …") in Example 12 (b), 0.02 and 0.009 are written as fractions so the calculation can be done mentally. Students may be confused about why 0.009 is written $\frac{0.9}{100}$. Explain that this is the same as dividing both the numerator and denominator by 10.

$0.009 = \frac{9}{1,000} = \frac{9 \div 10}{1,000 \div 10} = \frac{0.9}{100}$

We could also just write 0.009 as $\frac{9}{1,000}$ and multiply.

$3 \times 0.02 \times 0.009 = 3 \times \frac{2}{100} \times \frac{9}{1,000} =$

$\frac{3 \times 2 \times 9}{100 \times 1,000} = \frac{54}{100,000} = 0.00054$

- For Try It! 12, it may be helpful for students to make a sketch of the cuboid and write in the dimensions.

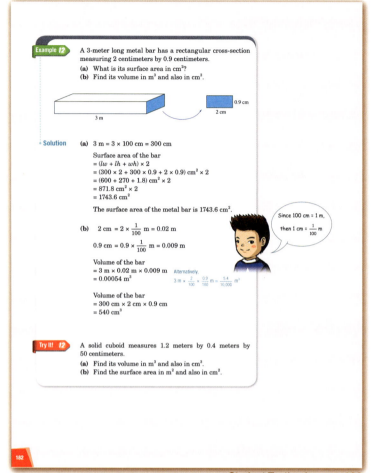

Student Textbook page 182

Try It! 12   Answers

(a) $50 \text{ cm} = 50 \times \dfrac{1}{100} \text{ m} = \dfrac{50}{100} \text{ m} = 0.5 \text{ m}$

Volume of cuboid $= 1.2 \text{ m} \times 0.4 \text{ m} \times 0.5 \text{ m} =$
0.24 m^3

$0.24 \text{ m}^3 = (0.24 \times 1{,}000{,}000) \text{ cm}^3 = 240{,}000 \text{ cm}^3$

Alternatively, to find the volume in cm^3, convert the length and width to cm first.

$1.2 \text{ m} = 120 \text{ cm}$

$0.4 \text{ m} = 40 \text{ cm}$

Volume of cuboid $= 120 \text{ cm} \times 40 \text{ cm} \times 50 \text{ cm} =$
$240{,}000 \text{ cm}^3$

(b) Surface area of cuboid
$= (1.2 \times 0.4 + 1.2 \times 0.5 + 0.4 \times 0.5) \text{ m}^2 \times 2$
$= 2.56 \text{ m}^2$

$2.56 \text{ m}^2 = (2.56 \times 10{,}000) \text{ cm}^2 = 25{,}600 \text{ cm}^2$

Alternatively, convert the length and width to cm first.

Surface area of cuboid $=$
$(120 \times 40 + 120 \times 50 + 40 \times 50) \text{ cm}^2 \times 2$
$= 12{,}800 \text{ cm}^2 \times 2$
$= 25{,}600 \text{ cm}^2$

Note: The answer for Try It! 12 (b) on p. 253 of the textbook should say $2.56 \text{ m}^2 = 25{,}600 \text{ cm}^2$ (not $2{,}56 \text{m}^2 = 25{,}600 \text{ cm}^2$).

4. Extension

Have students study Example 13 and do Try It! 13.

Notes:

Example 13:

- Since we do not need to find the area of the top base, we can modify the formula.
  Area to be tiled = Area of bottom base + (Area of front face + Area of side face) × 2

- For Try It! 13 (a), it may be helpful for students to make a sketch of the tank and write in the dimensions.

- For Try It 13 (b), the cover will go over the top and sides of the tank only so we are finding the surface area of one base and all of the sides.

Try It! 13   Answers

(a) 1 ft = 12 in

Capacity of tank = $(30 \times 8\frac{1}{3} \times 12)$ in³ = 3,000 in³

(b) Area to be covered =

Area of bottom base + Area of front face × 2 + Area of side face × 2

$= (30 \text{ in} \times 8\frac{1}{3} \text{ in}) + (30 \text{ in} \times 12 \text{ in} \times 2) +$

$(8\frac{1}{3} \text{ in} \times 12 \text{ in} \times 2)$

$= (250 + 720 + 200)$ in² = 1,170 in²

No, it is not large enough. It would have to be at least 1,170 in².

Note: In real life, dimensions of the plastic sheet would also have to be considered in order to cover the tank.

5. Conclusion

Summarize the main points of the lesson.
- Just as we can convert between linear units (e.g., m to cm), we can also convert between square units (e.g., m^2 to cm^2) and between cubic units (e.g., m^3 to cm^3).
- We can do this by multiplying or dividing by the conversion factor. For example:

  $1\ m^2 = 100\ cm^2$, thus $5\ m^2 = (5 \times 100)\ cm^2$

  $1\ cm^2 = \frac{1}{100}\ m^2$, thus $5\ cm^2 = \left(5 \times \frac{1}{100}\right)\ m^2$,

  or $5\ cm^2 = (5 \div 100)\ m^2$

- We can apply these ideas to solve real-life problems involving volume and surface area, even when the dimensions are given in different units, by converting to the same unit.

★ **Workbook: Page 122**

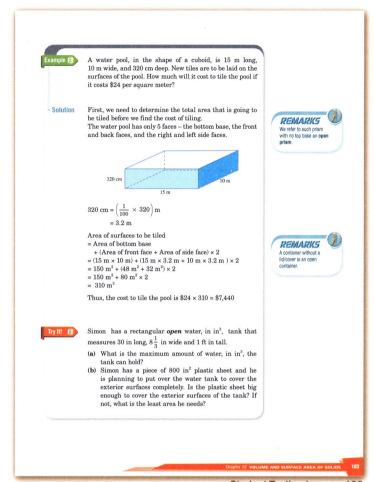

Student Textbook page 183

Lesson 8

Objective: Consolidate and extend the material covered thus far.

Have students work together with a partner or in groups. Students should try to solve the problems by themselves first, then compare solutions with their partner or group. If they are confused, they can discuss together.

Observe students carefully as they work on the problems. Give help as needed individually or in small groups.

BASIC PRACTICE

1. (a) 472 cm^2

   (b) 216 m^2
   The prism in (b) is a cube.

2. (a) 108 m^2 (b) 408 cm^2

3. (a) 64,000 cm^2 (b) 230,000 cm^2

   (c) 0.05 m^2 (d) 7.89 m^2

   (e) 42 mm^2 (f) 13.5 cm^2

FURTHER PRACTICE

4. (a) 0.456 m^3 (b) 0.0209 m^3

   (c) 8,000,000 cm^3 (d) 70,000 cm^3

   (e) 0.094 cm^3 (f) 1,060 mm^3

5. (a) Surface area = $97\frac{3}{4}$ in^2, volume = 55 in^3

   (b) Surface area = 734 cm^2, volume = 1,140 cm^3

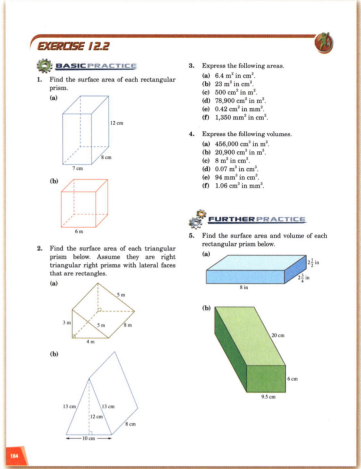

Student Textbook page 184

6. (a) Surface area = $16\frac{2}{3}$ in², volume = $4\frac{17}{27}$ in³

   (b) Surface area = $73\frac{1}{2}$ ft², volume = $42\frac{7}{8}$ ft³

7. (a) 360 m²

   (b) $13\frac{1}{2}$ ft²

8. (a) Triangular prism A, 19.8 cm²

MATH@WORK

9. (a) Surface area of cuboid = (3 × 2 + 3 × 3 + 3 × 2) ft² × 2
   $\quad\quad\quad\quad\quad\quad\quad\quad\quad\quad$ = 42 ft²

   $42 \div 18 = 2\frac{1}{3}$

   She cannot buy $2\frac{1}{3}$ cans, so she will need 3 cans.

   (b) $8.59 × 3 = $25.77

10. (a)
    Lateral surface area = (2.5 + 2.5 + 4) m × 8 m = 72 m²

    Combined area of bases = $\left(\frac{1}{2} \times 4 \times 1.5\right)$ m × 2 = 6 m²

    Total surface area of tent = 72 m² + 6 m² = 78 m²

    (b) $2.85 × 78 = $222.30

Student Textbook page 185

11. (a) (60 × 45 × 22) cm² + (20 × 20 × 20) cm³
                    = 59,400 cm³ + 8,000 cm³
                    = 67,400 cm³

    67,400 cm³ = $\left(67{,}400 \times \dfrac{1}{1{,}000{,}000}\right)$ m³ = 0.0674 m³

    (b) Note: The bottom face of the cube and the area of the cuboid covered by it are not part of the surface area of the solid.

    Total surface area of solid = (Surface area of cuboid + Surface area of cube) − (bottom face of cube × 2)

    Surface area of cuboid =
    (60 × 22 + 60 × 45 + 45 × 22) cm² × 2 =
       10,020 cm²
    Surface area of cube =
       (20 × 20) cm² × 6 = 2,400 cm²
    Surface area of solid =
       10,020 cm² + 2,400 cm² − (20 × 20 × 2) cm² =
       11,620 cm²

    11,620 cm² = $\left(11{,}620 \times \dfrac{1}{10{,}000}\right)$ m² = 1.162 m²

12. Surface area of 5 faces ⟶ 125 cm²
      Surface area of 1 face ⟶ 25 cm²
      5 × 5 = 25, so the length of one edge = 5 cm

    Or

    $(e \times e) \times 5 = 125$

    $e \times e = \dfrac{125}{5}$

    $e \times e = 25$

    $e = 5$

    Note: Students do not know square roots yet, but they should know that 5 × 5 = 25.

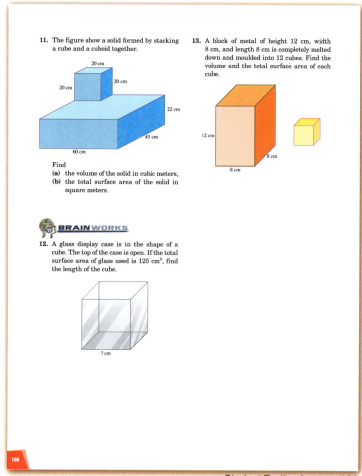

Student Textbook page 186

13. Volume of metal block =
       12 cm × 8 cm × 8 cm = 768 cm³
    Volume of one cube = 768 cm³ ÷ 12 = 64 cm³

    Edge × edge × edge = 64 cm³

    4 × 4 × 4 = 64

    Thus, the length of one edge = 4 cm
    Surface area of cube = (4 × 4) cm × 6 = 96 cm²

    Note: If students are having difficulty thinking of what number multiplied by itself 3 times is equal to 64, have them try 2 × 2 × 2, then 3 × 3 × 3, etc., until they get 64.

236

Lesson 9

Objective: Summarize and reflect on important ideas learned in this chapter, and solve a non-routine problem.

Note: This lesson could be done in class or assigned for students to do independently at home or in school.

1. In a Nutshell

Use this page to summarize the important ideas learned in this chapter.

Give examples where needed.

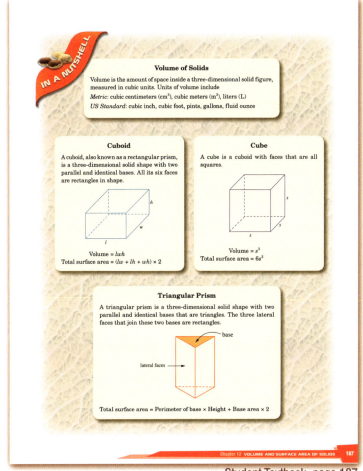

Student Textbook page 187

2. Write in Your Journal

Have students complete the writing activity and share their answers. Answers will vary.

(a) A prism has 2 bases that are polygons and lateral faces are all rectangles. A pyramid has one base that is a polygon and the other faces are all triangles.

(b) The bases of triangular pyramids and triangular prisms are triangles. However, the lateral faces of a triangular prism are rectangles, while the faces of a triangular pyramid are triangles.

3. Extend Your Learning Curve

This activity could be completed in class or done as an independent assignment.

Give students small square papers (such as origami paper or sticky notes).

Have students explore the possible nets of a cube by taping the sides of the square papers together to make a net. Then, they can see if their nets will form a cube.

Have students share the nets they made and discuss.

Notes:
- There are 11 possible nets of a cube.
- Encourage students to check to see if the nets they make are different or the same as the nets of their classmates. The net is different only if it cannot be flipped or rotated to make another net.
- Discuss the nets that do not work and ask students why they will not form a cube.

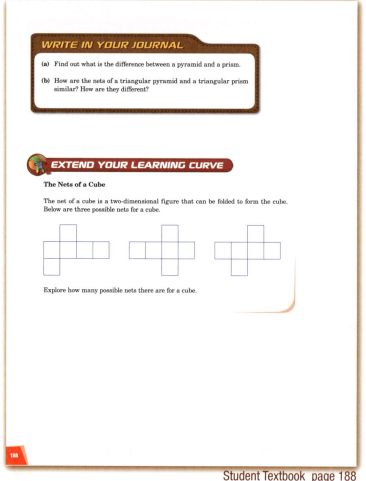

Student Textbook page 188

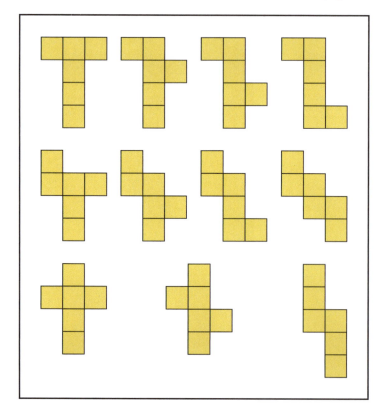

Lesson 10

Objective: Solve complex and non-routine problems involving solid figures.

1. Problem Solving

Have students study Examples 1 – 2 and do Try It! 1 – 2. Students can work individually or in pairs or groups. If students are working in pairs or groups they should try the problems independently first and then share and discuss with their groups.

Have students share and discuss their solutions and methods.

Note: If students are having difficulty, encourage them to draw the two triangular bases at opposite ends first, and then draw the middle rectangular face.

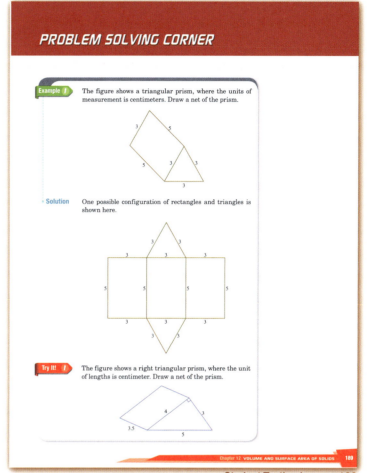

Student Textbook page 189

Try It! 1 Answers

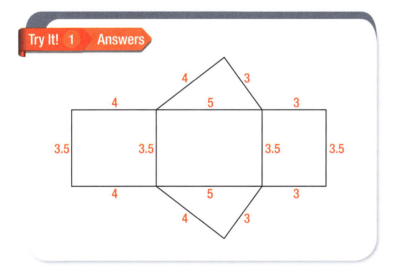

Chapter 12 VOLUME AND SURFACE AREA OF SOLIDS 239

Try It! 2 Answers

Amount of water in tank after being filled for 5 minutes =

$$10 \text{ L} \times 5 = 50 \text{ L}$$

Amount of water in tank after being drained for 6 minutes =

$$7 \text{ L} \times 6 = 42 \text{ L}$$

Amount of water remaining in the tank =

$$50 \text{ L} - 42 \text{ L} = 8 \text{ L} = 8,000 \text{ mL} = 8,000 \text{ cm}^3$$

Length × width × height = Volume

$$50 \text{ cm} \times 50 \text{ cm} \times h = 8,000 \text{ cm}^3$$

$$\text{Height} = \frac{8,000 \text{ cm}^3}{50 \text{ cm} \times 50 \text{ cm}} = 3.2 \text{ cm}$$

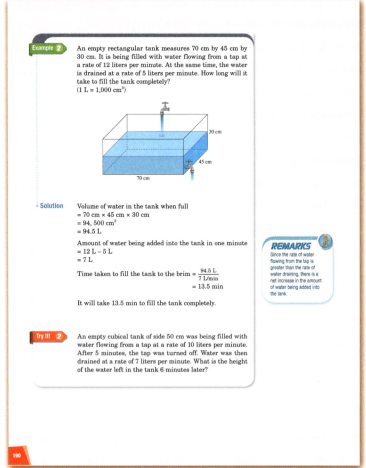

Student Textbook page 190

2. Practice

Have students do Practice Questions 1 – 10.

Students can work together and help each other but they should try the problems independently first.

1. No. of $\frac{1}{3}$-unit cubes that can fit in a 1-unit cube = 3 × 3 × 3 = 27

   No. of $\frac{1}{3}$-unit cubes that can fit in a prism with capacity of 6 cubic units = 27 × 6 = 162

   No. of $\frac{1}{4}$-unit cubes that can fit in a 1 cubic unit prism = 4 × 4 × 4 = 64

   No. of $\frac{1}{4}$-unit cubes that can fit in a prism with capacity of 6 cubic units = 64 × 6 = 384

   384 − 162 = 222

   It takes 222 more $\frac{1}{4}$-unit cubes than $\frac{1}{3}$-unit cubes to fill the prism.

2.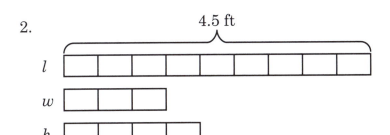

   Length (9 units) → 4.5 ft
   1 unit → 4.5 ft ÷ 9 = 0.5 ft
   Width (3 units) → 0.5 ft × 3 = 1.5 ft
   Height (5 units) → 0.5 ft × 4 = 2 ft

   Volume of prism = 4.5 ft × 1.5 ft × 2 ft = 13.5 ft³

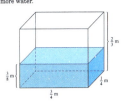

Student Textbook page 191

3. (a) Capacity of tank = $\frac{3}{4}$ m × $\frac{1}{4}$ m × $\frac{2}{3}$ m = $\frac{1}{8}$ m³

   (b) Difference in height = $\frac{2}{3}$ m − $\frac{3}{8}$ m = $\frac{7}{24}$ m

   Amount of water needed to fill the tank =

   $\frac{3}{4}$ m × $\frac{1}{4}$ m × $\frac{7}{24}$ m = $\frac{7}{128}$ m³

   Note: Alternatively, students can subtract the volume of the water in the tank from the capacity of the tank.

   Volume of water in the tank =

   $\frac{3}{8}$ m × $\frac{3}{4}$ m × $\frac{1}{4}$ m = $\frac{9}{128}$ m³

   $\frac{1}{8}$ m³ − $\frac{9}{128}$ m³ = $\frac{7}{128}$ m³

4. Surface area of box =
   $(40 \times 30 + 40 \times 5 + 30 \times 5)$ cm² $\times 2 = 3{,}100$ cm²
   Area of paper = 80 cm × 40 cm = 3,200 cm²

   3,200 cm² > 3,100 cm², so yes, there is enough paper.

5. (a) Capacity of present box =
   7 in × 3 in × 10 in = 210 in³
   Capacity of new box =
   9 in × 9 in × 3 in = 243 in³
   The new box will hold more cookies.

   (b) Assuming the boxes will have a top.
   Surface area of present box =
   $(7 \times 3 + 7 \times 10 + 3 \times 10)$ in² $\times 2 = 242$ in²
   Surface area of new box =
   $(9 \times 9 + 9 \times 3 + 9 \times 3)$ in² $\times 2 = 270$ in²
   The new box will require more material.

6. Volume of water in Container P =
   9 cm × 6 cm × 12 cm = 648 cm³
   Length × width × height =
   Volume of water in Container Q
   12 cm × 10 cm × height = 648 cm³
   Height = $\frac{648 \text{ cm}^3}{12 \text{ cm} \times 10 \text{ cm}}$ = 5.4 cm

7. Volume of water = 1.5 L = 1,500 mL = 1,500 cm³
   Length of tank = $\frac{1{,}500 \text{ cm}^3}{9 \text{ cm} \times 6 \text{ cm}} = \frac{1{,}500}{54}$ cm

   Volume of oil = 2.5 L = 2,500 mL = 2,500 cm³
   Height of oil = 2,500 cm³ ÷ $\left(\frac{1{,}500}{54} \text{ cm} \times 9 \text{ cm}\right)$
   = 2500 cm³ ÷ 250 cm = 10 cm

8. Volume of water in Container P =
   30 cm × 25 cm × 50 cm = 37,500 cm³

   $37{,}500 = 40 \times 15 \times h + (20 \times 20 \times h)$
   $37{,}500 = 600h + 400h$
   $37{,}500 = 1{,}000h$
   $h = \frac{37{,}500}{1{,}000}$
   $h = 37.5$

   Note: The answer given on page 254 says 37.5 cm. However, the unit is not given here so the answer can be expressed as 37.5 or 37.5 units.

9. (a) Area of face A = 16 cm × 16 cm = 256 cm²

   Ratio of the area of face A to the area of face B = 4 : 5

   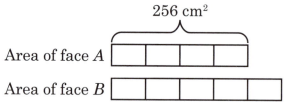

   Area of face A (4 units) → 256 cm²
   1 unit → 256 cm² ÷ 4 = 64 cm²
   Area of face A (5 units) → 64 cm² × 5 = 320 cm²

   Length of longer side of face B
   = $\frac{320 \text{ cm}^3}{16 \text{ cm}}$ = 20 cm
   = width of cuboid

   Length of cuboid = 16 cm
   Width of cuboid = 20 cm
   Height of cuboid = 16 cm

   (b) 16 cm × 20 cm × 16 cm = 5,120 cm³

10. **Note:** Have students think about the values first and then trace and cut out the net to form a cube to check their answers.

    X is opposite of 6, so $X = 1$.
    Y is opposite of 2, so $Y = 5$.
    Z is opposite of 3, so $Z = 4$.

---

7. A rectangular tank is 9 cm wide. It contains 1.5 L of water and 2.5 L of oil. The height of the water level is the tank is 6 cm. What is the height of the oil level?

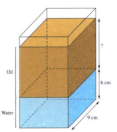

8. Rectangular container P is completely filled with water. All the water is then poured into two empty rectangular containers, Q and R, so that the height of the water is the same in both containers. What is the height of the water in each container?

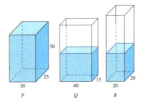

9. The figure below shows a cuboid with a square face A and a rectangular face B. The ratio of the area of A to the area of B is $4 : 5$.

   (a) What are the dimensions of the cuboid?
   (b) Find the volume of the cuboid.

10. On a die, the values on the opposite faces add up to 7. The diagram below shows a piece of cardboard that can be folded to form a die. How many dots are on the faces, X, Y, Z?

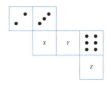

Student Textbook page 192

Notes

Chapter 13: Displaying and Comparing Data

Lesson	Objectives	Class Periods	Textbook & Workbook	Teacher's Guide Page	Additional Materials Needed
1	• Recognize a statistical question. • Find the mean of a set of data.	1	TB:193–196	249	
2	• Find the median and mode of a set of data.	1	TB:197–200	253	
3	• Collect and analyze data. • Determine which measure of center best represents a set of data.	1	TB:201–203 WB:126–133	257	
4	• Consolidate and extend the material covered thus far.	1	TB:204–206	260	
5	• Determine whether a given number makes an inequality true.	1	TB:207–212 WB:134–136	265	Grid paper, rulers
6	• Use a frequency table and a histogram to display data grouped in intervals. • Describe the shape of a set of data.	1	TB:213–220 WB:137–144	271	Grid paper, rulers
7	• Consolidate and extend the material covered thus far.	1	TB:220–224	279	
8	• Use range to analyze variability in a set of data.	1	TB:224–226 WB:145–146	283	
9	• Find the Mean Absolute Deviation of a set of data.	1	TB:226–227	286	
10	• Solve real-life problems involving Mean Absolute Deviation. • Determine whether the mean is a good indicator of the typical values in a data set.	1	TB:228–231 WB:147–150	288	

Continues on next page.

©2017 Singapore Math Inc. Dimensions Math® Teacher's Guide 6B

Chapter 13: Displaying and Comparing Data

Lesson	Objectives	Class Periods	Textbook & Workbook	Teacher's Guide Page	Additional Materials Needed
11	• Use interquartile range to determine variability around the median.	1	TB: 231–233 WB: 151–153	292	
12	• Use a box plot to summarize a data distribution.	1	TB: 234–240 WB: 154–155	296	Grid paper, rulers
13	• Consolidate and extend the material covered thus far.	1	TB: 241–243	303	
14	• Summarize and reflect on important ideas learned in this chapter, and solve a non-routine problem.	1	TB: 244–246	307	

Statistics involves collecting, organizing, displaying, analyzing, and interpreting data. Two important types of statistical measures are **measures of center** and **measures of variability**. Each of these measures gives us different information about the data set.

Measures of center tell us the typical value of a data set and are often called **averages**. Three common averages are **mean**, **median**, and **mode**.

The **mean** of a data set is the sum of all the data values divided by the total number of data values.

$$\text{Mean} = \frac{\text{Sum of the values in a data set}}{\text{Number of values in a data set}}$$

The **median** of a data set is the middle value when the data values are arranged from smallest to largest. The median is often a better indicator of the typical value in a data set than the mean when there are extreme values that may skew (distort) the mean.

24   24   25   26   58   60   60

↑

median age
(in the "middle position")

The ages of 7 people in a group are shown. The age of person who is 60 years old skews the mean to make it seem like the typical person in the group is older. The median, 26, is a better indicator of the typical age than the mean, which is 31.

The **mode** is the value that occurs most often in a data set.

Heights of Basketball Players (cm)

178	175	168	170	178	165
172	175	170	175	175	

The modal height of the basketball players is 175 cm. There can be more than one mode in a set of data.

Chapter 13: Displaying and Comparing Data

Measures of center do not give us the whole picture of the values in a data set. Two data sets can have the same mean but the values within each set could vary considerably. Variability in a distribution refers to how close or far apart the values within a data set are. In **Dimensions Math® 6**, students will learn about three measures of variability — **range**, **mean absolute deviation**, and **interquartile range**.

The **range** of a data set is the difference between the largest and smallest values. Range can be affected by extreme values in the data, so other measures of variability are helpful to get a better picture of how the data varies.

Mean absolute deviation (MAD) is the average distance of the data values from the mean. A small MAD means that data values are clustered closely. When the data values are clustered around the mean, we say the distribution of the data has **low variability**. A large MAD means that the data values are spread apart. When the data values are spread out away from the mean, we say that the distribution of the data has **high variability**.

The **interquartile range** is the range of the middle 50% of the data. The interquartile range eliminates the lower 25% and the upper 25% of the data which negates the effect of the outliers and gives a better picture of how the typical data in a set varies.

It is easier to see how data is distributed within a set if we organize the values in tables and graphs. In this chapter, students will learn to display data in **dot plots**, **histograms**, and **box plots**.

A **dot plot** is useful to highlight clusters and gaps in small sets of data, as well as values that vary considerably from the center.

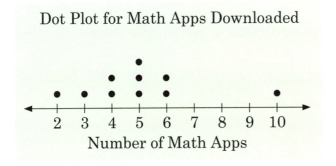

In this dot plot, we can see that the data values are clustered around 5. The value 10 is an extreme value that varies significantly from the center.

Chapter 13: Displaying and Comparing Data

Histograms are used for displaying larger sets of data because the data is grouped in intervals. A skewed distribution has values that are not typical of the rest of the data. By looking at the shape of the histogram, we can tell whether or not the values are skewed toward the higher or lower values.

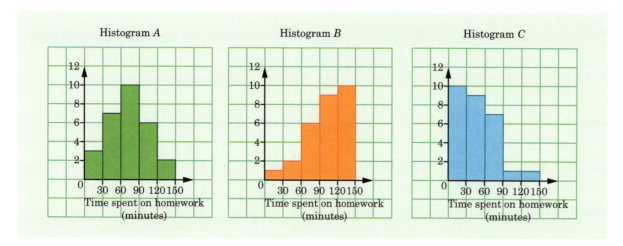

Histogram A is a mound shape so the data is not skewed significantly. Histogram B is skewed left toward the smaller values, i.e., the smaller values make is seem like the average value is smaller than is really typical for the data. Histogram C is skewed right toward the larger values, i.e. the larger values skew the mean to make is seem like the average value is larger than is really typical.

A box plot is useful to see the spread of data more clearly by dividing it into 4 equal sections that each contain 25% of the data. A box plot can be used for large data sets. It is also called a **box and whisker plot**.

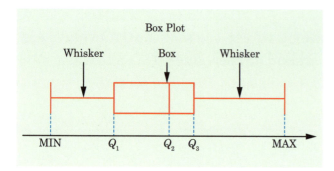

A box plot shows the range of the lower 25% of the data, the range of the middle 50% of the data (interquartile range), and the range of the upper 25% of the data. It also gives us a 5-point summary of the variability of the data, which includes the lowest value (MIN), the median of the lower half of the data (Q_1), the median of the data (Q_2), the median of the upper half of the data (Q_3), and the highest value (MAX).

Lesson 1

Objectives:
- Recognize a statistical question.
- Find the mean of a set of data.

1. Introduction

Read and discuss page 193, and state the learning objectives of the chapter, LET'S LEARN TO …

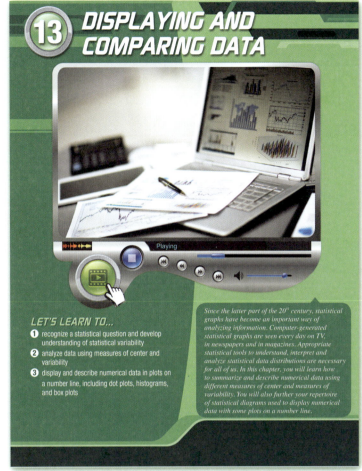

Student Textbook page 193

2. Development

Have students read the top of page 194 and discuss the criteria for a statistical question. Have students study Example 1, do Try It! 1, and discuss the solutions.

Try It! 1 Answers

(a) This is a statistical question because the heights of the students vary.

(b) This is not a statistical question. The answer would not be expected to vary, so it has only one answer.

(c) This is not a statistical question. The answer would not be expected to vary, so it has only one answer.

(d) This is a statistical question, because the months students have their birthdays vary.

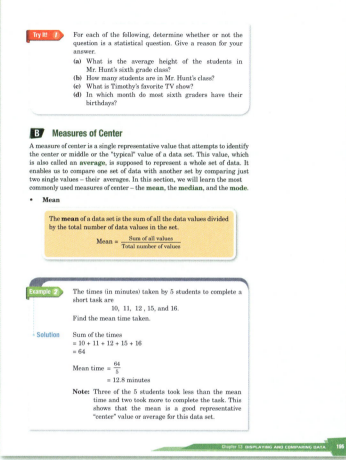

Student Textbook page 195

3. **Application**

Read and discuss the middle of page 195.

Have students study Example 2 and do Try It! 2.

Notes:
- "What is the mean age of 5 students?" is a statistical question, because we would need to collect data that varies about the ages of the student.
- Students learned about average in **Dimensions Math® 6A** Chapter 6 as a way to even out several quantities by dividing the total amount by the number of items. They were actually finding the mean, which is the arithmetic average. (See REMARKS on page 196.) Students will learn about other types of averages (median and mode) in the next lesson.

> **Try It! 2 Answer**
>
> $$\frac{21 + 25 + 18 + 20 + 22 + 17 + 23 + 18}{8} = \frac{164}{8} = 20.5$$
>
> Mean age = 20.5 years

4. Extension

Have students study Example 3 and do Try It! 3.

Notes:

Example 3:
- When we multiply the mean by the total number of values, we get the sum of all the values in the data set.

  Mean = $\frac{\text{Sum of values}}{\text{Number of values}}$, thus **Mean ×**

  Number of values = Sum of values

- This can also be solved algebraically.

  $\frac{28 + 35 + x}{3} = 34$

  $28 + 35 + x = 34 \times 3$

  $63 + x = 102$

  $\quad\quad x = 102 - 63$

  $\quad\quad x = 39$

Try It! 3 involves using two averages to find the combined ages of the parents and the combined ages of the parents plus Mila's age. Mila's age is the difference between these two amounts.

> **Try It! 3 Answers**
>
> (a) Sum of 4 numbers = 20.5 × 4 = 82
>
> x = 82 − (13 + 23 + 29) = 17
>
> (b) Combined ages of parents = 42 × 8 = 336
>
>   Combined ages of parents and Mila
>   = 40 × 9 = 360
>
>   Mila's age = 360 − 336 = 24
>
>   Mila is 24 years old.

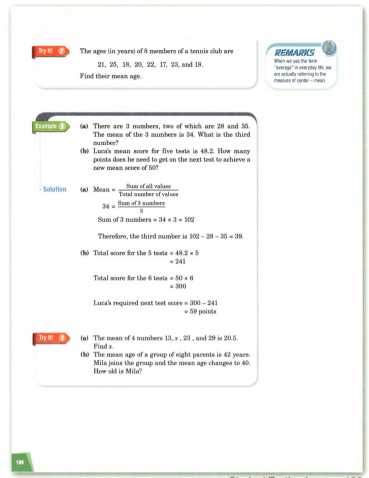

Student Textbook page 196

5. Conclusion

- A statistical question is a question that requires using data that varies to find the answer.
- We can summarize the data we collect by using single values that are representative of the whole data set. The mean is one way of summarizing how the data clusters.

252

©2017 Singapore Math Inc. Dimensions Math® Teacher's Guide 6

Lesson 2

Objective: Find the median and mode of a set of data.

1. Introduction

Pose the problem on the top of page 197. (Have students cover the answer with an index card or paper). Ask, "Is 31 a good indicator of the typical age of a teacher in this group? Why or why not?"

Note: Help students see that 5 of the 7 teachers are under 30, so 31 may not be a good indicator of the typical age of the teachers in the group.

Have students arrange the ages of the teachers in ascending order and find the middle value (median age).

Read and discuss page 197.

Notes:

- Since one teacher is 60 years old, this extreme value skews (distorts) the mean toward the higher values so it appears that the average age of a teacher is older than typical of the group. The mean age is 31 years, but we can see that most of the teachers are in their twenties, so the median is a better representation of the typical age of this group.
- Extreme values will only skew the mean if they are weighted heavily on one side (i.e. far away from most of the data). For example, if there were a 1-year-old included in the data the median and mean would be closer to each other.

　　1, 24, 24, 25, 26, 28, 30, 60
　　Mean = 218 ÷ 8 ≈ 27
　　Median = (25 + 26) ÷ 2 = 25.5

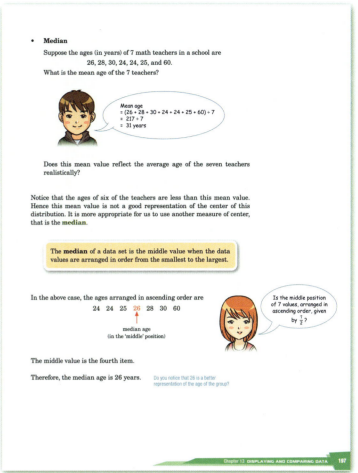

Student Textbook page 197

In this case, the effect of the high extreme value is mostly negated by the effect of the low extreme value. An extreme value that skews the data is known as an outlier. (See RECALL on page 208.)

2. Application

Have students study Example 4 and do Try It! 4.

Example 4:
- There is an even number of data points, so the median is the mean of the two middle numbers. The position of the median can be determined visually. Alternatively, we can find the middle position by adding 1 to the number of data points and dividing by 2.

The middle position = $\frac{n+1}{2}$, where n = the number of data points. Since there are 8 data points, the position of the median can be determined by $\frac{8+1}{2}$ = 4.5, which means that the median is halfway between the 4th and 5th values. This method is good for locating the median in a large data set.

Try It! 4 Answer

There is an even number of data points (already arranged in order), so the median is the mean of the two middle data points.

$\frac{14.5 + 15}{2}$ = 14.75

Median height ≈ 14.8 cm

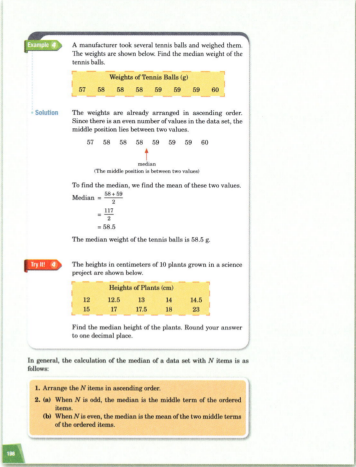

Student Textbook page 198

Use the bottom of page 198 and the top of page 199 to summarize this part of the lesson.

Have students talk about DISCUSS with a partner or in groups.
- Regardless of whether the data is arranged in ascending or descending order, the middle values will be the same.

3. Development

Pose the problem about the shoes in the middle of page 199 (cover the answer) and ask students to find the mean and median.

Ask, "Do the mean and median provide useful information for the manufacturer of the shoes and the store owner? Why or why not?"
- The mean and median give shoe sizes that do not exist. A better typical value of the shoe sizes would be the most popular shoe size (the mode).

Have students read the bottom half of page 199 with a partner or in groups and discuss how to find the mode of a set of data.

Discuss the bottom of page 199.

Notes:
- If the shoe sizes sold were 7, 8, 9, 10, 11, and 12, there would be no mode because none of the data occur more often. (See REMARKS.)

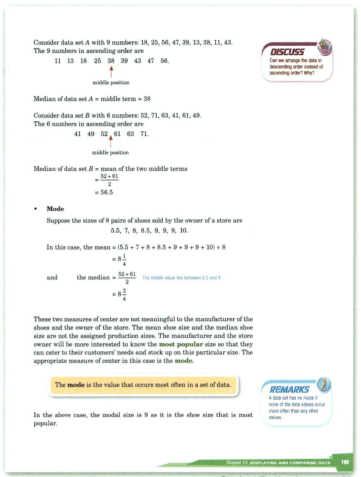
Student Textbook page 199

4. Extension

Have students study Example 5 and do Try It! 5.

Notes:

Example 5:
- A set of data can have more than one mode. For Quiz *B*, there are two modal scores because two scores occur the same number of times.

Use the bottom of page 200 to summarize this part of the lesson.
- Categorical data is data that can be divided into categories or groups; e.g., age group, ethnicity, level of education, etc.
- The mode is not always a number. It can be a category (e.g., "tennis").

5. Conclusion

Summarize the main points of the lesson.
- Mean and median are useful averages when we have numerical data.
- When there are extreme values that skew the data toward the higher or lower values, median is generally a better representation than mean.
- For categorical data, mode is useful because the frequency a data point occurs is an important consideration for this type of data.

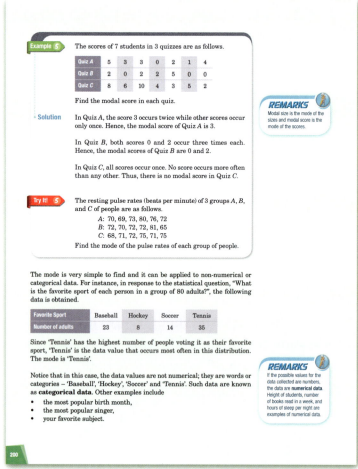

Student Textbook page 200

A: No mode

B: 72 beats per minute

C: 71 and 75 per minute

Lesson 3

Objectives:
- Collect and analyze data.
- Determine which measure of center best represents a set of data.

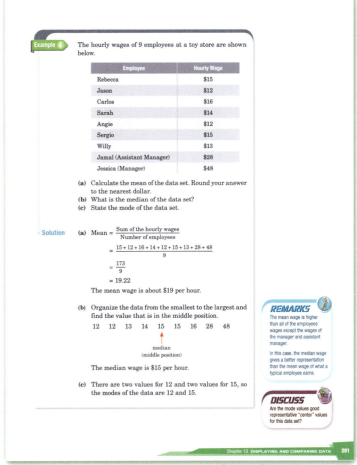

Student Textbook page 201

1. Introduction

Remind students how to find mean, median, and mode.

Have students study Example 6 and do Try It! 6.
- It may be helpful to allow students to use calculators for these problems.

Have students talk about REMARKS and DISCUSS with a partner or in groups.

Notes:
- In these examples, students are comparing the three measures of central tendency they have learned. The median is a better indication of the typical wage of an employee, because the two highest wages skew the data toward an average wage that is higher than typical.
- The mode tells us the most common wages paid to employees, but it does not represent the center of the data well, because $12 per hour is the lowest wage. However, if there were many more employees making $12 per hour than all other wages, $12 per hour would be a good indicator of the typical wage at this store.

Try It! 6   Answers

(a) (178 + 175 + 168 + 170 + 178 + 165 + 172 + 175 + 170 + 175 + 175) cm ÷ 11 =
    1,901 cm ÷ 11 ≈ 172.8 cm ≈ 173 cm

(b) 165, 168, 170, 170, 172, **175**, 175, 175, 175, 178, 178
    Median height = 175 cm

(c) There are four 175-cm players, making 175 cm the most common height. Modal height = 175 cm.

2. Development

Have students work together with a partner to complete Class Activity 1.

Notes:
- Students can include themselves in the survey.
- Students' conclusions will vary. If there are extreme values in the data, then the median may be a better choice. For example, if one or a few students spend much more time on homework than everyone else, median would be a better indicator.

Have students share the results of their surveys and their conclusions.

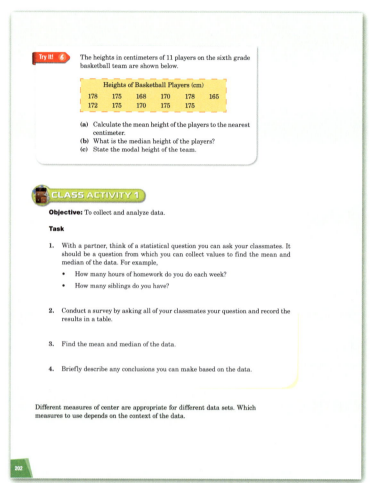

Student Textbook page 202

3. **Application**

Have students complete Class Activity 2.

Answers for Class Activity 2

1. (a) Mode (Size 6 is most common.)

   (b) Median (The highest wage skews the mean.)

   (c) Mean (There are no extreme values.)

2. (a)

Measure of Center	Mean	Median	Mode
Data set A	2.5	1.5	1
Data set B	7	4.5	15
Data set C	5.2	3	1

   (b) Mean

   (c) Mean

   Note: Some students looking at the table may think that the mode was the most affected. If so, explain that the mode would not be affected by large or small values because it is the value that occurs most often. The reason the mode changed in (c) is because now 1 is the most common data point, not 15.

3. Use the bottom of page 203 to summarize the activity.

4. **Conclusion**

Summarize the main points of the lesson.
- Mean and median are generally the most useful measure of center for numerical data.
- When there are extreme values that skew the data, median may be better than mean.
- Mode is useful when we have categorical data.

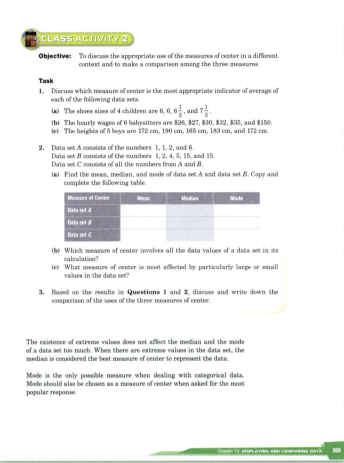

Student Textbook page 203

★ **Workbook: Page 126**

Lesson 4

Objective: Consolidate and extend the material covered thus far.

Have students work together with a partner or in groups. Students should try to solve the problems by themselves first, then compare solutions with their partner or group. If they are confused, they can discuss together.

Observe students carefully as they work on the problems. Give help as needed individually or in small groups.

Notes:
- Students may use calculators at the teacher's discretion. They should show their work.

1. (a) Yes, because you need to collect information that varies on people's income.

    (b) No, because there is only one answer and no variability.

    (c) No, because there is only one answer and no variability.

    (d) Yes, because you need to collect data that varies on students' favorite ice cream.

2. (a) Mode, because hair color is categorical data.

    (b) Mean, usually, because it is numerical data and usually the extreme values will not skew the data.

    (c) Median, usually, because it is numerical data and there could be tall or short plants that skew the data.

    (d) Mode, because coat size is categorical data.

    (e) Mean, usually, because it is numerical data and usually the extreme values will not skew the data.

Student Textbook page 204

Note:
- For 2 (b), (c), and (e), the answer could be either mean or median so there could be different answers. It would depend on whether there were extreme values that could skew the data.

3. (a)    (b) Blue and red

    (c) Pear

4. (a) (i) Mean: 7, Median: 8, Mode: 10

    (ii) Mean: 5, Median: 3, Mode: 3

    (iii) Mean: 6.5, Median: 6, Mode: 5

    (iv) Mean: 9, Median: 9.5, Mode: no mode

    (v) Mean: 8.3, Median: 6, Mode: 4

(b) (i) Mean, because there are no extreme values.
 (ii) Mean, because there are no extreme values.
 (iii) Mean, because there are no extreme values.
 (iv) Mean, because there are no extreme values.
 (v) Median, because there is an extreme value, 33.

FURTHER PRACTICE

5. (a) $50 \times 10 = 500$

 (b) $42.5 \times 10 = 425$
 $425 - 74 = 351$
 $351 \div 9 = 39$
 39 years old

6. (a) 36, 54, 75, 85, 85, 87, 90, 92

 Median $= \dfrac{85 + 85}{2} = 85$

 (b) $(87 + 92 + 54 + 85 + 90 + 75 + 85 + 36) \div 8 =$
 $604 \div 8 = 75.5$

 (c) Median $= \dfrac{85 + 87}{2} = 86$

 Mean $= (604 - 36 + 96) \div 8 = 664 \div 8 = 83$
 54, 75, 85, 85, 87, 90, 92, 96

 (d) The mean will be affected because the mean is more easily skewed by extreme values. The median will change, but probably not that much because extreme values do not usually affect the median as much.

7. $1{,}800 - (350 \text{ mi} \times 4) = 1{,}800 - 1{,}400 = 400$
 He has to drive 400 miles.

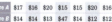

Student Textbook page 205

8. (a) $x + 2 = 11$
 $x = 9$

 (b) $x + 2 = 7$
 $x = 5$

 Note: If 9 is the median and the difference between the highest and smallest numbers is 4, then the smallest number is $11 - 4 = 7$, thus $x + 2 = 7$, $x = 5$.

 (c) $\dfrac{9 + (x + 2) + 11}{3} = 12$

 $9 + (x + 2) + 11 = 12 \times 3$
 $x + 22 = 36$
 $x = 36 - 22$
 $x = 14$

9. Note: The answers given on textbook page 255 in the first printing are wrong. The correct answers are:

    (a) (19.3 + 10.5 + 24 + 11.6 + 13.6 + 16 + 13.5 + 19.5 + 38.7 + 12 + 15.2 + 12.5) ÷ 12 = 206.4 ÷ 12
    = 17.2 lb

    (b) 10.5, 11.6, 12, 12.5, 13.5, 13.6, 15.2, 16, 19.3, 19.5, 24, 38.7

    $\dfrac{13.6 + 15.2}{2}$ = 14.4 lb

    (c) The median because 38.7 is an extreme value that skews the mean.

10. (a) Player A
    Mean = (18 + 12 + 22 + 21 + 19 + 24) ÷ 6 =
    116 ÷ 6 ≈ 19.3 points

    Median = $\dfrac{19 + 21}{2}$ = 20 points

    Player B
    Mean = (42 + 4 + 12 + 10 + 40 + 10) ÷ 6 =
    118 ÷ 6 ≈ 19.7 points

    Median = $\dfrac{10 + 12}{2}$ = 11 points

    (b) Player A → 24 − 12 = 12 points
    Player B → 42 − 4 = 38 points

    (c) Player A is more consistent so the coach should probably choose Player A.
    Note: Player B could be chosen if the coach is looking for a player who can score many points in one game. The choice depends on the coach's criteria for choosing a player.

Student Textbook page 206

11. Store A:
    Mean wage =
    (17 + 16 + 20 + 15 + 15 + 20 + 16) ÷ 7 = $17
    15, 15, 16, 16, 17, 20, 20, Median wage = $16

    Store B:
    Mean wage =
    (14 + 14 + 13 + 47 + 14 + 12 + 12) ÷ 7 = $18
    12, 12, 13, 14, 14, 14, 47, Median wage = $14

    (a) Fiorella owns Store A because it has a higher median wage.

    (b) Monica owns Store B because it has a higher mean wage.

    Note: Ask the students whose employees they think are really happier and why.

12. Students' examples will vary.

    (a) False   (b) False

    (c) False   (d) False

13. (a) $1 + 2 + 4 + 3 + + 4 + 5 + 2 = 21$ students

    (b) 5 children have a score of 19, so the modal score is 19.

    (c)

Mark	14	15	16	17	18	19	20
Number of Students	1	2	4	3	4	5	2

    (d) List the values in ascending order by writing one 14, two 15s, four 16s, etc., and then find the middle value.
    14, 15, 15, 16, 16, 16, 16, 17, 17, 17, 18, 18, 18, 18, 19, 19, 19 ,19, 19, 20, 20

    (e) The modal score is higher than most of the scores, so the median score is a better representative of the data.

14. (a) $(2 + 5 + 8 + 6 + 9) \div 5 = 30 \div 5 = 6$

    (b) $(12 + 15 + 18 + 16 + 19) \div 5 = 80 \div 5 = 16$
    If you add 10 to each data point, the mean will also be 10 more.
    Note: The mean is the second to last number in both data sets, only because the data sets have the numbers in the same order they did before, and because it just so happened it was that number in the first data set. Its position is not important, except insofar as they are keeping the order of the data points the same, and that number happened to correspond to the mean. The principle here has nothing whatsoever to do with being a number in the data set. But since it is, if the order is kept the same, it will be in the same position in the new data set. Now, if one understands this, and sees that the order is being kept the same in all of them, and in all of them the same number is being added to each data point, then you can use that pattern, but if the data set were 2, 5, 8, 9, and 11, the mean is 7, and not in the data set. Add 10 to each, 12, 15, 18, 19, 21, the mean will be 7 + 10, 17, also not in the data set.

    2, 5, 8, 6, 9   12, 15, 18, 16, 19

    (c) We can use the mean of the data set in (a) to find the means of the other data sets.
    In (i), 30 was added to each value of the data set in (a) so instead of a mean of 6 the mean is $6 + 30 = 36$.
    In (ii), 220 was added to each value of the data set in (a) so the mean is $6 + 220 = 226$.
    In (iii), 3 was added to each value of the data set in (a), so mean is $6 + 3 = 9$.
    In (iv), each value of the data set in (a) was doubled, so the mean is $6 \times 2 = 12$.

15. $$p + q + r = 9 \times 3 = 27$$
    $$(p + q + r) + (s + t) = 15 \times 5 = 75$$
    $$27 + (s + t) = 75$$
    $$s + t = 75 - 27$$
    $$s + t = 48$$
    Mean of s and $t = \dfrac{s + t}{2} = \dfrac{48}{2} = 24$

16. (a) Students can list all the data values, excluding the unknown, and then use a logical trial and error method with a calculator.

$\dfrac{35 + 2}{22 + 1} \approx 1.6$, $\dfrac{35 + 4}{22 + 2} \approx 1.625$ ... etc.

It goes up slowly to 1.7 with each trial, so they can skip a few to narrow in on one that gives an answer of $\dfrac{35 + 16}{22 + 8}$, so $x = 8$.

It can also be solved algebraically.

$$\dfrac{\text{Total number of siblings}}{\text{Number of students}}, = \text{Mean}$$

$$\dfrac{(5 \times 0) + (9 \times 1) + (x \times 2) + (6 \times 3) + (4 \times 2)}{5 + 9 + x + 6 + 2}$$

$$= 1.7$$

$$\dfrac{2x + 35}{22 + x} = 1.7$$

$$2x + 35 = 1.7(22 + x)$$

$$2x + 35 = 37.4 + 1.7x$$

$$2x - 1.7x = 37.4 - 35$$

$$0.3x = 2.4$$

$$x = 2.4 \div 0.3$$

$$x = 8$$

(b) If we write out all of the data points without the median, we can see that there are 14 values to the left of the median.

0, 0, 0, 0, 0, 1, 1, 1, 1, 1, 1, 1, 1, 1, . . . , 3, 3, 3, 3, 3, 3, 4, 4

There has to be 14 values to the right of the median. There are already 8 values on the right, so you would need six 2s to the right of the median and one 2 in the middle (as the median), so $x = 7$.

0, 0, 0, 0, 0, 1, 1, 1, 1, 1, 1, 1, 1, 1, 2, 2, 2, 2, 2, 2, 2, 3, 3, 3, 3, 3, 3, 4, 4

(c) Since 9 students have 1 sibling, it would have to be at least 10.

Lesson 5

Objective: Use a dot plot to display numerical data.

1. Introduction

Read the top of page 207.

Give students grid paper and rulers and have them follow the steps on page 207 to construct a dot plot of the data.

Ask, "What can we see about the data by looking at the dot plot?"

Have students discuss the answer with a partner or in groups and share.
- Answers should be similar to the top of page 208.

Student Textbook page 207

Read and discuss the top of page 208.

2. Development

Have students study Example 7 and do Try It! 7.

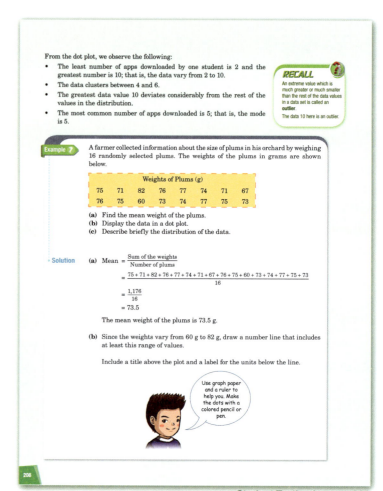

Student Textbook page 208

Notes:

- Students may use calculators at the teachers' discretion.
- In statistics, a distribution is an arrangement that shows the frequency of each value in a data set. By looking at the distribution, we can see how the values in the data set are spread out and where they cluster.
- In a set of data, half of a data set are to the left of the median and half the values are to the right of the median. Thus, the median is found where the data are clustered, not necessarily in the middle of the distribution.

Try It! 7 Answers

(a) (57 + 54 + 56 + 60 + 58 + 55 + 55 + 50 + 56 + 61 + 55 + 56 + 68) ÷ 13 = 741 ÷ 13 = 57 °F

Note: In the first printing of the textbook, the answer given for Try It! 7 (a) on page 254 is wrong.

(b)
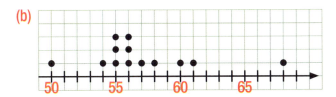

(c) From the dot plot, we observe the following:
- The data vary from 50 °F to 68 °F.
- The data cluster between 54 °F and 58 °F.
- The lowest temperature and the highest temperature deviate considerably from the mean temperature.

Note: The first printing of the textbook states incorrectly the data cluster is between 54 °F and 61 °F as part of the answer to (c).

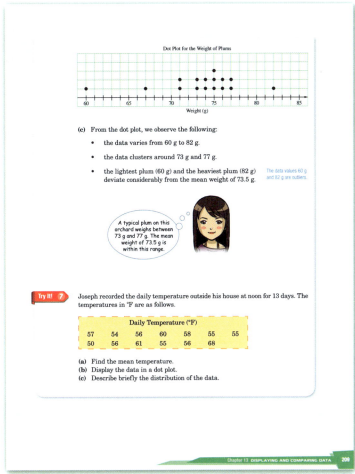

Student Textbook page 209

3. Application

Have students study Example 8 and do Try It! 8.

Notes:

Example 8:

- See the second REMARKS for a simple way to find the median by crossing off dots from each end. This would be the same as listing all the values in ascending order and crossing off values from each end until you get to the middle value (or middle two values if there is an odd number of values).

- Another way to find the position of median is to add 1 to the total number of dots and divide by 2. $\frac{20 + 1}{2} = 10.5$, so the median is halfway between the 10th and 11th values. Counting off dots starting from the top left, the 10th and 11th dots are both 2s. The median is the mean of these two values, thus median = $\frac{2 + 2}{2} = 2$.

Dot Plot for Number of Children

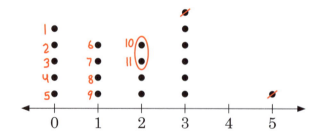

Number of Children in Each Family

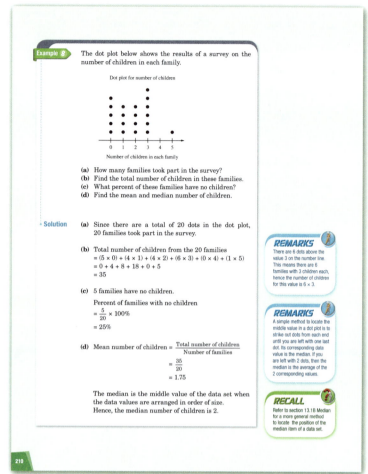

Student Textbook page 210

Try It! 8 Answers

(a) 20

(b) 4 × 1 + 6 × 2 + 4 × 3 + 2 × 4 + 4 × 5 = 4 + 12 + 12 + 8 + 20 = 56

(c) $\frac{6}{20} \times 100\% = 30\%$

(d) Mean = $\frac{56}{20}$ = 2.8

Median = $\frac{2+3}{2}$ = 2.5

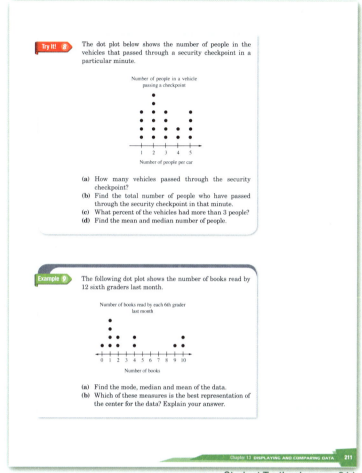

Student Textbook page 211

4. **Extension**

Have students study Example 9 and do Try It! 9.

Notes:

Example 9:
- In these problems, students must determine which measure of center best describes the data by looking at the distribution of the data.
- The mode would not be a good representation of the data because most students did not read 1 book.
- The mean is skewed toward the higher values by the students who read 9 and 10 books. Thus, the median is a better representation of the data. When there are outliers, the median is usually a better representation than the mean.
- For Try It! 9 (b), the mean is a better representation because there are no extreme values that skew the data, and the data is distributed fairly evenly. When the data distribution is fairly even, the values of the median and the mean will be close.

Try It! 9 Answers

(a) Median: 5 books
    Mean: 5.4 books

(b) The mean because there are no extreme values.

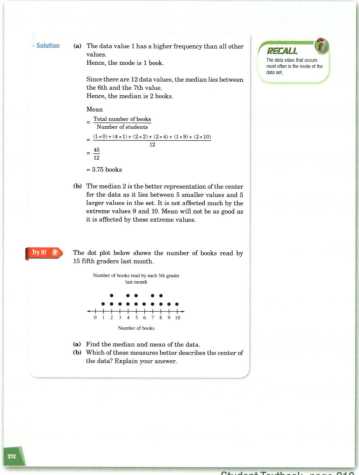

Student Textbook page 212

5. **Conclusion**

Summarize the main points of the lesson.
- We can use a dot plot to display small sets of data. We can use the dot plot to find the mean, median, and mode of the data.
- By looking at the dot plot, we can see how the values in the data set are distributed. This helps us determine which measure of center best represents the data.

★ **Workbook: Page 134**

Lesson 6

Objectives:
- Use a frequency table and a histogram to display data grouped in intervals.
- Describe the shape of a set of data.

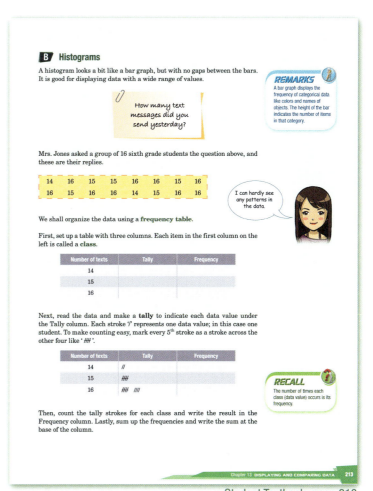

Student Textbook page 213

1. Introduction

Read pages 213 – 214 and discuss how to organize the data in a frequency table.

Notes:
- When students learned about bar graphs in elementary school, they mostly used them for categorical data. Numerical data often have many data points that are spread out across a range of values. Frequency tables and histograms allow us to organize numerical data into groups to see patterns in the data more easily.
- When the number of values is small, it is often difficult to see patterns in the data. (See what the girl is saying on page 213.) When there are more data values, more patterns in the data emerge. (See table on page 214.)

- In the second survey, more students were surveyed. When there are many data values, we can organize the data into groups called **classes** with a range of values called an **interval**.
- The **span** (size) of each interval is chosen based on the number of values so there are not too many or too few classes to compare. The interval 0 – 4 has 5 values, 0, 1, 2, 3, and 4, so the span of the interval is 5. Each interval should have the same span so the comparison is valid and the bars of the histogram have the same width. Discuss the guidelines for choosing class intervals at the bottom of page 214.

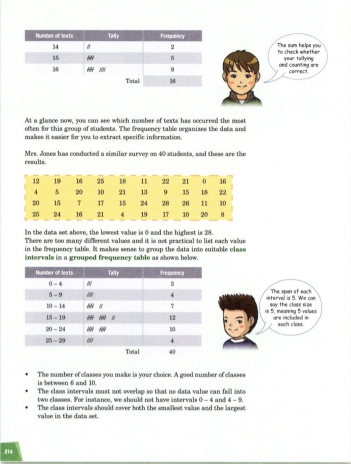

Student Textbook page 214

2. Development

Give students grid paper and rulers and have them follow the steps given to draw a histogram for the data from Mrs. Jones' second survey given in the frequency table.

Discuss the differences between the bar graph and the histogram on the bottom of page 215.

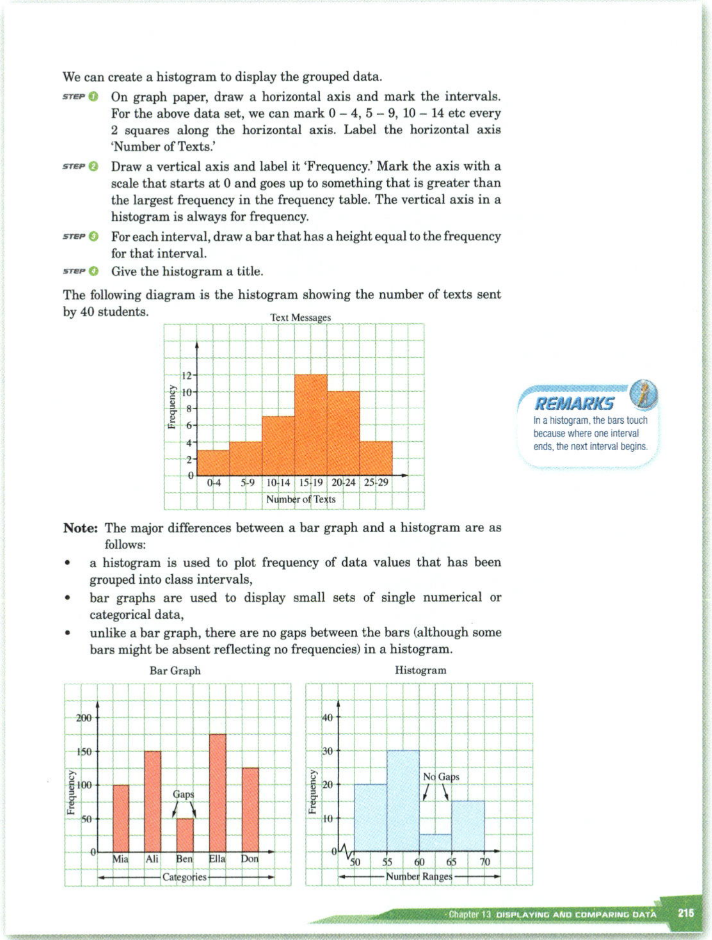

We can create a histogram to display the grouped data.

STEP ❶ On graph paper, draw a horizontal axis and mark the intervals. For the above data set, we can mark 0 – 4, 5 – 9, 10 – 14 etc every 2 squares along the horizontal axis. Label the horizontal axis 'Number of Texts.'

STEP ❷ Draw a vertical axis and label it 'Frequency.' Mark the axis with a scale that starts at 0 and goes up to something that is greater than the largest frequency in the frequency table. The vertical axis in a histogram is always for frequency.

STEP ❸ For each interval, draw a bar that has a height equal to the frequency for that interval.

STEP ❹ Give the histogram a title.

The following diagram is the histogram showing the number of texts sent by 40 students.

REMARKS
In a histogram, the bars touch because where one interval ends, the next interval begins.

Note: The major differences between a bar graph and a histogram are as follows:

- a histogram is used to plot frequency of data values that has been grouped into class intervals,
- bar graphs are used to display small sets of single numerical or categorical data,
- unlike a bar graph, there are no gaps between the bars (although some bars might be absent reflecting no frequencies) in a histogram.

Student Textbook page 215

Chapter 13 DISPLAYING AND COMPARING DATA **273**

3. **Application**

Read and discuss the top of page 216.

Notes:

- Discrete quantities are quantities that do not exist in fractional amounts (e.g., number of texts, number of people, etc.). Continuous quantities are quantities that exist in fractional amounts (e.g. meters, minutes, etc.). When continuous quantities are involved we need to have class intervals that take into account fractional amounts, so we show the class interval using inequality signs.
- Since we do not want any data values to overlap, the inequality sign on the left of the variable is $\leq$ and the inequality sign on the right side of the variable is <. Since it is an inequality, the span of the interval includes an infinite number of values.

Have students study Example 10 and do Try It! 10.

Example 10:

- For 10 (a), the interval $9 \leq x < 12$ includes 9, but not 12. The interval includes all values from 9 to any number less than 12. The next interval must include 12 but exclude 15 so it is $12 \leq x < 15$.
- Point out to students that the intervals need to be **uniform** (have the same span). To check for uniformity, make sure the difference between the upper and lower limits of each interval are the same. For example, $12 - 9 = 3$, $15 - 12 = 3$, $18 - 15 = 3$, etc.

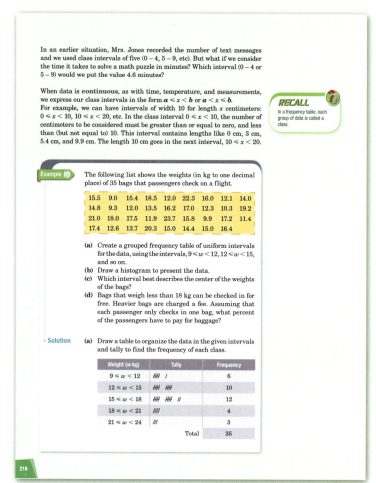

Student Textbook page 216

Have students talk about the REMARK and DISCUSS questions with a partner or in groups.

- The "$-\bigwedge-$" on the x-axis helps us to save space on the horizontal axis when creating the graph. (See REMARK.)
- When the data is distributed fairly evenly, the mean and median are close to each other and found in the center of the distribution. Thus, $15 \leq w < 18$ best describes the center. (See DISCUSS 1).
- The histogram and dot plot both show the spread and shape of the data but the dot plot shows individual data values while the histogram shows groups of values. The histogram is better for a large data set because by creating groups we can include many more values. This would be impractical on a dot plot. (See DISCUSS 2.)

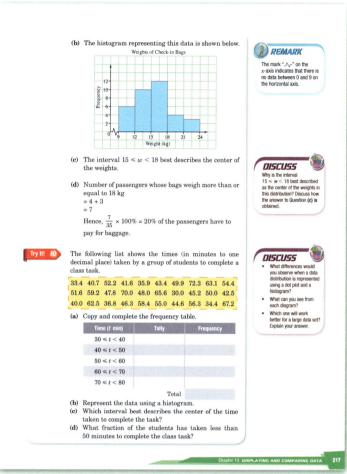

Student Textbook page 217

Try It! 10 Answers

(a)

Time (t min)	Tally	Frequency
$30 \leq t < 40$	|||||	5
$40 \leq t < 50$	||||| |||||		11
$50 \leq t < 60$	||||| |||	8
$60 \leq t < 70$	||||	4
$70 \leq t < 80$	||	2
	Total	30

(b)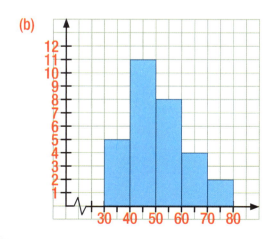

(c) Interval $40 \leq t < 50$

(d) $\frac{16}{30} = \frac{8}{15}$

4. Extension

Read and discuss page 218.

Notes:

- Students learned about line symmetry in **Dimensions Math® 4**. Remind them that a line of symmetry divides a shape into two sides that are mirror images of each other.
- When the distribution is almost symmetric, the data is clustered in the middle and there are no outliers that skew the mean (i.e., if there are outliers, they are distributed evenly on both sides of the mean so they basically cancel each other out). The mean and median will both be close to the middle. The mean is usually the best measure of center.
- When the distribution is skewed left, the mean will be skewed by the smaller values on the left. The data is clustered on the right, so the median is closer to the right side. The median is usually the best measure of center.
- When the distribution is skewed right, the mean will be skewed by the larger values on the right. The data is clustered on the left, so the median will be closer to the left side. The median is usually the best measure of center.
- When a distribution is even (i.e., all the bars are the same height) we say that it has a **uniform** distribution. Uniform distributions are also considered to be symmetric. The mean is usually the best measure of center.

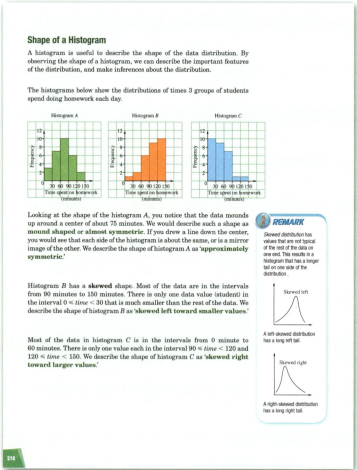

Student Textbook page 218

Have students study Example 11.

Have students discuss the conclusions in Example 11 (d) with a partner or in groups.

- For Try It! 11 (b), the data is in Histogram *B* clustered left so the median is closer to the left side of the distribution. The interval $60 \leq$ length < 70 best represents the center because approximately half of the values are on each side of this interval.

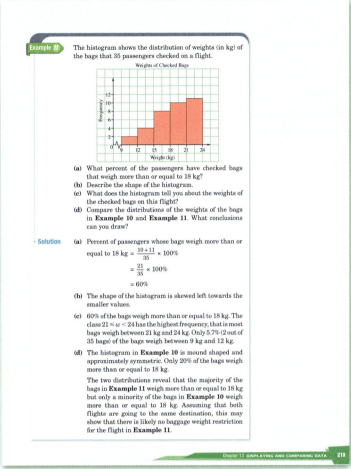

Student Textbook page 219

Have students do Try It! 11 and discuss the answers.

Try It! 11 Answers

(a) Histogram A: approximately symmetric
Histogram B: skewed right

(b) Histogram A:
the interval 70 cm $\leq$ length < 80 cm
Histogram B:
the interval 60 cm $\leq$ length < 70 cm

(c) Histogram A: class 70 cm $\leq$ length < 80 cm with 25 ropes
Histogram B: class 50 cm $\leq$ length < 60 cm with 25 ropes

(d) For A, the shape of the histogram is approximately symmetric around a center of about 75 cm. Most of the ropes are between a length of 60 and 90 cm. About $\frac{1}{3}$ of the ropes are between the length of 70 cm and 80 cm.
For B, the shape of the histogram is skewed right. Most of the ropes (about $\frac{2}{3}$) are between 50 cm and 70 cm. There are very few ropes between 80 cm and 100 cm.

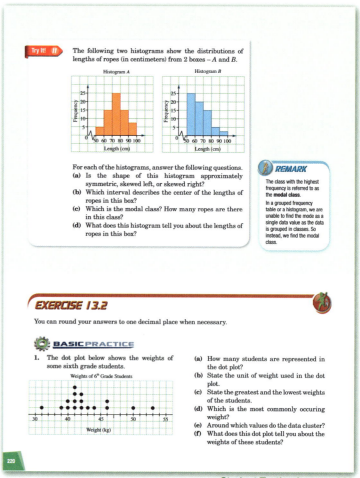

Student Textbook page 220

5. Conclusion

Summarize the main points of the lesson.

- We can use frequency tables and histograms to display larger sets of numerical data by dividing the values into groups (classes).
- When we have continuous quantities (quantities that can have fractional amounts), we show the intervals using the inequality signs ($\leq$ and <).
- By looking at the histogram, we can see the shape of the data and find its center. This can help us to draw conclusions about the data.

★ **Workbook: Page 137**

Lesson 7

Objective: Consolidate and extend the material covered thus far.

Have students work together with a partner or in groups. Students should try to solve the problems by themselves first, then compare solutions with their partner or group. If they are confused, they can discuss together.

Observe students carefully as they work on the problems. Give help as needed individually or in small groups.

Notes:
- Students may use calculators at the teacher's discretion.

1. (a) 17   (b) Kilogram

   (c) Greatest weight: 66 kg
   Lowest weight: 32 kg

   (d) 42 kg   (e) Between 38 kg and 48 kg

   (f) Most students weigh between 39 kg and 44 kg. Two students weigh a lot more and one student weighs a lot less.

   Note: The scale of the dot plot in the textbook is wrong. It should have been 30   40   50   60   70 instead of 30   40   45   50   55.

2. (a) 25   (b) 10 points

   (c) 21 to 30

   (d) Class 31 to 40

   (e) $\frac{12}{25} \times 100\% = 48\%$

   Note: the answer on textbook page 255 is given as a fraction, not a percentage.

Student Textbook page 221

3. (a)

Amount spent, x	Frequency
$5 \leq x < 15$	1
$15 \leq x < 25$	0
$25 \leq x < 35$	4
$35 \leq x < 45$	8
$45 \leq x < 55$	5
$55 \leq x < 65$	2

(b) $10

(c) No, $44.99 is in interval $35 \leq x < 45$, and $45 is in interval $45 \leq x < 55$.

(d) Class $35 \leq x < 45$

(e) The average customer spends between $35 and $45.

FURTHER PRACTICE

4. (a) Dot plot. The data is skewed right toward larger values.

   (b) Dot plot. The data is approximately symmetric.

   (c) Histogram. The data is skewed right toward larger values.

   (d) Histogram. The data is approximately symmetric.

5. (a) 118 apps

   (b) 5.9 apps

   (c) Median: 6 apps
       Mode: 5 apps

   (d) The distribution is approximately symmetric.

   (e) Mean. The distribution is approximately symmetric so the mean is the best representation of the center.

6. (a)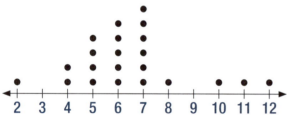
   Dot Plot for Total Number on Dice

   (b) Approximately symmetric

   (c) 6.5

   (d) 6

   (e) Mean, because the distribution is approximately symmetrical.

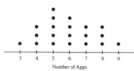

Student Textbook page 222

7. (a)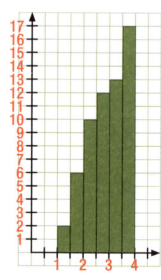

   (b) 50%

   (c) Skewed left

   (d) The mass of most of the parcels is between 2 kg and 4 kg. Only 8 parcels are less than 2 kg.

8. (a) $\frac{20 + 14 + 5}{3 + 6 + 12 + 20 + 14 + 5} \times 100\% = \frac{39}{60} \times 100\% = 65\%$

   (b) Class height between 160 cm to 165 cm.

   (c) It is approximately symmetric. Most of the girls are between 155 and 170 cm tall.

MATH@WORK

9. (a) Carton A is approximately symmetric. Most of the eggs have a mass of 60 g or 61 g. Carton B is skewed left toward smaller values. Most of the eggs have a mass from 60 g to 63 g.

   (b) Most of the eggs in Carton B have a greater mass than the eggs in Carton A.

10. (a)

Distance, x, in meters	Frequency
$3 \leq x < 4$	3
$4 \leq x < 5$	5
$5 \leq x < 6$	9
$6 \leq x < 7$	11
$7 \leq x < 8$	4

   (b)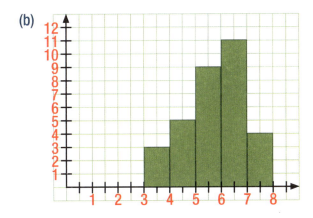

   (c) Interval $5 \leq x < 6$

   (d) 8 boys

11. (a) Boys $\rightarrow \dfrac{18 + 15}{18 + 15 + 12 + 5} \times 100\% = \dfrac{33}{50} \times 100\% = 66\%$

   Girls $\rightarrow \dfrac{14 + 12}{14 + 12 + 12 + 12} \times 100\% = \dfrac{26}{50} \times 100\% = 52\%$

   (b) The distribution of the boys is skewed right toward the higher values. The distribution of the girls is mostly even, but skewed slightly right toward the higher values.

   (c) 34% of the boys and 48% of the girls are older than 10 years old, so the girls in the school are mostly older than the boys.

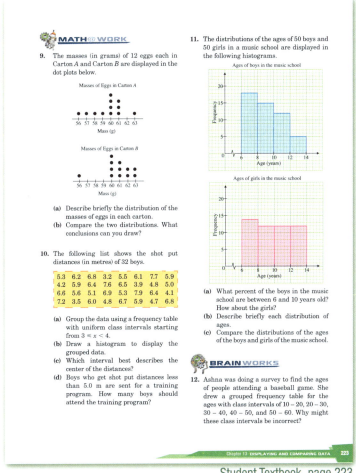

Student Textbook page 223

12. There is overlap in the intervals. A person who is 20 years old, for example, would be represented in two classes, 10 – 20 and 20 – 30.

13. Note: In these class intervals, the lower value is not included, but the upper value is included, so the intervals are written in the form $a < x \leq b$ (not $a \leq x < b$).

(a) (i)

Amount Raised, x	Frequency
$0 < x \leq 4$	4
$4 < x \leq 8$	9
$8 < x \leq 12$	5
$12 < x \leq 16$	4
$16 < x \leq 20$	5
$20 < x \leq 24$	3

(ii)

Amount Raised, x	Frequency
$0 < x \leq 6$	9
$6 < x \leq 12$	9
$12 < x \leq 18$	6
$18 < x \leq 24$	6

(b) (i)

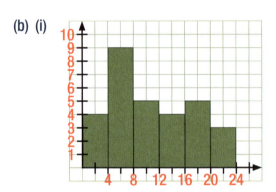

(ii)

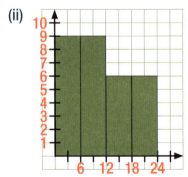

(c) 1. The modal class is $4 < x \leq 8$.
   2. The modal class is $0 < x \leq 6$ and $6 < x \leq 12$.
      Note: The answers given on textbook page 256 of the first printing are wrong.

(d) She should select the one with the class size of $4, because it shows the shape of the distribution better.

Lesson 8

Objective: Use range to analyze variability in a set of data.

1. Introduction

Ask students to find the mean of each of the following sets of data.
A: 16, 17, 18, 19, 20
B: 7, 14, 18, 22, 29

Note: This is the same problem as the problem on page 224. The calculations for the mean and median are shown in the margin in blue.

Ask, "What do you notice about the mean and median?" (They are the same.)

Ask, "What do you think the distribution will look like?"

Possible responses:
- They will both be symmetric.
- They might look similar.
- It looks like A might be more spread out (wider).

Have students make dot plots for each set of data.

Tell students to use the same scale for each line in order to compare the dot plots. (See below.)

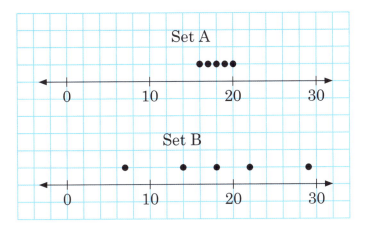

Ask, "What do you notice about how the data values are spread out? Why do you think they are spread out this way?"
- B is more spread out than A, because in A the values are clustered in the middle.
- B is more spread out than A because the lowest and highest values are much different.
- In A there isn't much difference in the values so they are close together.

Read and discuss page 224.

2. Development

Have students study Example 12, do Try It! 12, and share and discuss their solutions.

Ask, "What does the range tell us about the spread (variability) of the data?"
- The greater the range, the more spread out the values are.
- The smaller the range, the more clustered the values are.

Try It! 12 Answers

(a) 41 − 9 = 32

(b) 50 − 6 = 44

3. Application

Have students do Example 13 (cover the solution).

Ask them to find the mean and range for each students' test scores and share their answers.

Ask students to create a dot plot for each students' scores and to draw an arrow to indicate the location of the mean on each dot plot.

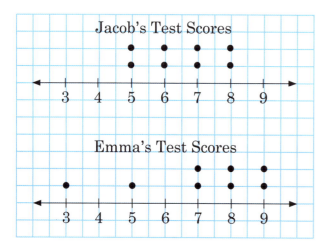

Ask, "What conclusions can you draw about Jacob's and Emma's results based on the means, ranges and dot plots?"
- Emma's results are better because her mean score is higher.
- Emma's results are very different because she had a 3 on one test and a 9 on another test.
- Jacob's scores are clustered closer together so his scores are similar. His scores are more consistent.

Ask, "Whose scores are closer to the mean score?" (Jacob's)

Ask, "What can the range tell us about their scores that the mean does not tell us?"
- The mean gives us the average score but it doesn't tell how different the scores are.
- We can conclude other things about the data by looking at the range, such as how consistent each student's scores are.

Discuss the solution given on page 225.

4. **Extension**

Have students do Try It! 13 and share and discuss their solutions.

Try It! 13   Answers

(a) Mean temperature for City $A =$

$= \dfrac{49 + 50 + 47 + 45 + 42 + 40 + 42}{7} = \dfrac{315}{7} = 45 \,°F$

Range for City $A = 50°F - 40°F = 10°F$

Mean temperature for City $B =$

$= \dfrac{50 + 51 + 52 + 52 + 48 + 46 + 46}{7} = \dfrac{345}{7} = 49.3 \,°F$

Range for City $B = 52°F - 46°F = 6°F$

(b) The weather in City B is generally warmer than in City A as it has a higher mean temperature. City A has a larger range, this tells us that the temperature in City A is less consistent than in City B.

Note: In the first printing of the textbook, the answer in the back of the book for (b) is wrong.

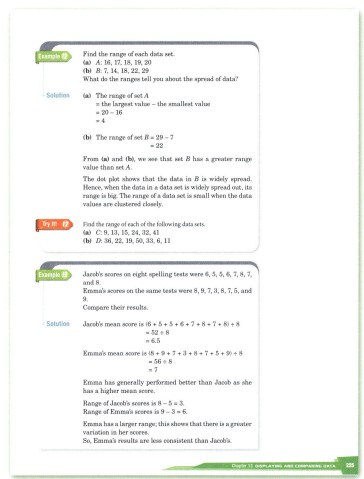

Student Textbook page 225

5. **Conclusion**

Summarize the main points of the lesson.
- Variability refers to how spread out the values in the data set are. Range is a measure of variability.
- By looking at variability, we can draw conclusions about the data that cannot be drawn by just looking at measures of center.

★ **Workbook: Page 145**

Lesson 9

Objective: Find the mean absolute deviation of a set of data.

1. Introduction

Ask students to find the mean and make dot plots for each of the following sets of data:

Data Set *A*: 3, 3, 5, 5
Data Set *B*: 1, 1, 6, 8

Note: These are the same sets of data given on page 226.

Ask, "For which data set are the data values closer to the mean?" (Data set *B*)

Ask, "How far is each data value in Set *A* from the mean?"

- 3 → 1 unit
  3 → 1 unit
  5 → 1 unit
  5 → 1 unit

Ask, "What is the mean distance of each data value in Set *A* from the mean?"

- $\frac{1+1+1+1}{4} = 1$

Ask, "How far is each data value in Set *B* from the mean?"

- 1 → 3 units
  1 → 3 units
  6 → 2 units
  8 → 4 units

Ask, "What is the mean distance of each data value in Set *A* from the mean?"

- $\frac{3+3+2+4}{4} = 3$

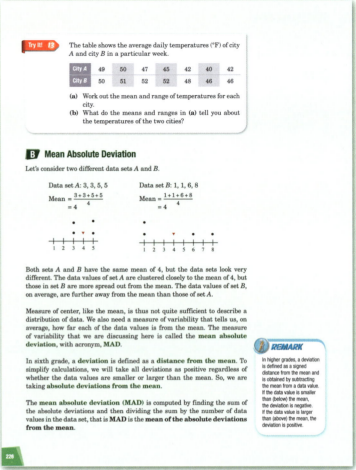

Student Textbook page 226

2. Development

Ask, "What does the average distance of the data values from the mean tell us about the distribution of the data?" Possible responses:

- On average, the values in Set *A* are 1 away from the mean. On average, the values in Set *B* are 3 away from the mean.
- In Set *A*, the values are generally closer to the mean. In Set *B* the values are generally further away from the mean.
- The values in Set *A* are spread out more than the values in Set *B*.
- The greater the average distance from the mean, the more spread out the values are. The smaller the average distance from the mean, the less spread out the values are.

Read and discuss page 226.

3. Application

Use the top of page 227 to show students how to find the sum of the absolute deviations for each set of data using absolute value.

Note: Focus students on the diagrams with the arrows at the top of page 227. The number line can be used to count the absolute deviation of a data value from the mean. This is the same as finding the absolute value of the difference between the data value and the mean.

Discuss the procedure for finding Mean Absolute Deviation (MAD).

4. Extension

Have students discuss what the boy is saying with a partner or in groups. Read and discuss the summary box at the bottom of page 227.

Notes:
- If we consider the mean as 0, the numbers to the left are considered negative and the numbers to the right positive. For example, for Set B:

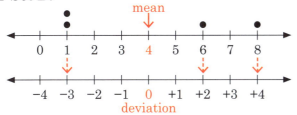

The deviation of the data point 1 is −3, which means it deviates 3 units from the mean in the negative direction. Its absolute deviation is $|-3| = 3$. The absolute deviation is the unsigned distance of the data value from the mean. (See RECALL.)

- A small MAD indicates low variability, which means that the values are close to the middle (i.e., close to the mean). A large MAD indicates high variability, which means that the values are more spread out (i.e., farther away from the mean).

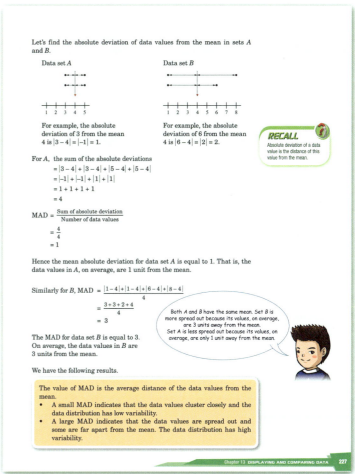

Student Textbook page 227

- Range is a limited measure of variability because it only tells us the difference between the largest and smallest values. If most the data values are clustered around the mean but one data point is farther away from the mean, the data set may have a large range but low variability. Thus, MAD gives us more accurate information about the variability of the data.

5. Conclusion

Summarize the main points of the lesson.
- The Mean Absolute Deviation (MAD) is the mean distance of the data values from the mean.
- MAD tells us about the variability in the data (i.e., how spread out or close together the data values are).

Lesson 10

Objectives:
- Solve real-life problems involving Mean Absolute Deviation.
- Determine whether the mean is a good indicator of the typical values in a data set.

1. Introduction

Remind students of the procedure for finding the Mean Absolute Deviation for a set of data.

Have students solve Example 14 (cover the answers).

Notes:
If students are having difficulty finding the absolute deviation of the values from the mean, encourage them to use a number line. (See REMARK.)

Have students share and discuss their solutions. Discuss the solution given on page 228.

Have students make a dot plot for the ages of the children and label the mean.

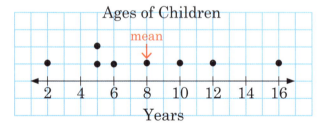

Ask, "What would the dot plot look like if there were a higher Mean Absolute Deviation?"

Possible responses:

- It would be more spread apart.
- There would be fewer values that are close to 8.
- There would more older or younger people.

Ask, "What would the dot plot look like if there were a lower Mean Absolute Deviation?"

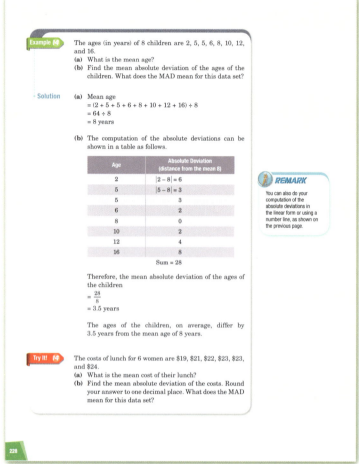

Student Textbook page 228

Possible responses:

- The values would be more clustered in the middle.
- There would be more values that are close to 8.
- There would fewer older or younger people.

2. Development

Have students do Try It! 14 and share and discuss their solutions.

Try It! 14 Answers

(a) Mean cost of lunch = $(19 + 21 + 22 + 23 + 23 + 23 + 24) ÷ 6 = $132 ÷ 6 = $22

(b) Sum of absolute deviations
= |19 − 22| + |21 − 22| + |22 − 22| + |23 − 22| + |23 − 22| + |24 − 22|
= 3 + 1 + 0 + 1 + 1 + 2
= 8

MAD = $\frac{8}{6}$ ≈ 1.3 ≈ $1.30

On average, the cost of lunch differs from the mean of $22 by $1.30.

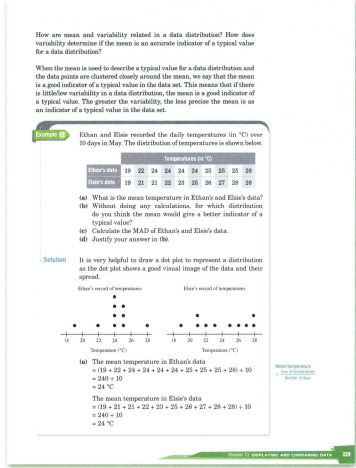

Student Textbook page 229

3. Application

Have students draw a dot plot and find the mean for each set of data in Example 15 (cover the answer).

Have students answer questions (a) – (d) and discuss their answers with partners or in groups. Then, ask them to share their conclusions.

Read and discuss the top of page 229.

Discuss the solutions given on page 229 – 230.

Summarize this part of the lesson by discussing the box at the bottom of page 230.

Note: Just because two sets of data have a similar mean it does not mean that the sets of data are similar. Mean Absolute Deviation helps us to compare two sets of data with similar means by looking at the variability within each data set.

4. Extension

Have students do Try It! 15 and share and discuss their solutions.

> **(b)** The dot plots show that most of the temperatures in Ethan's data are closely clustered to the mean of 24, while those in Elsie's data are more spread out.
>
> Thus Ethan's data has lower variability than Elsie's and the mean should be a better indicator of a typical value for Ethan's data distribution.
>
> **(c)** In Ethan's data,
> the sum of the absolute deviations from the mean
> $= 5 + 2 + 0 + 0 + 0 + 0 + 1 + 1 + 1 + 4$
> $= 14$
> MAD $= \frac{14}{10}$
> $= 1.4$
>
> In Elsie's data,
> the sum of the absolute deviations from the mean
> $= 5 + 3 + 3 + 2 + 1 + 1 + 2 + 3 + 4 + 4$
> $= 28$
> MAD $= \frac{28}{10}$
> $= 2.8$
>
> **(d)** The values of the MAD in **(c)** confirm the answer in **(b)**. The MAD of 1.4 for Ethan's data is much smaller than the MAD of 2.8 for Elsie's data.
>
> Hence the mean is a better indicator of a typical value for Ethan's data distribution than for Elsie's.
>
> In summary, we have the following.
>
> > The mean is a better indicator of a typical value in a data set when there is low variability in the distribution than when there is high variability.

Student Textbook page 230

Try It! 15   Answers

(a)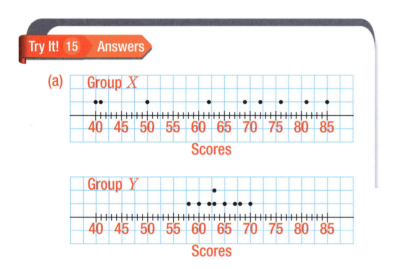
Group X

Group Y

(b) Mean for group $X = (40 + 41 + 50 + 62 + 69 + 72 + 76 + 81 + 85) \div 9 = 576 \div 9 = 64$

MAD for group $X =$

$= \dfrac{|40 - 64| + |41 - 64| + |50 - 64| + |62 - 64| + |69 - 64| + |72 - 64| + |76 - 64| + |81 - 64| + |85 - 64|}{9}$

$= \dfrac{24 + 23 + 14 + 2 + 5 + 8 + 12 + 17 + 21}{9} = \dfrac{126}{9} = 14$

Mean for group Y $= (58 + 60 + 62 + 63 + 63 + 65 + 67 + 68 + 70) \div 9 = 576 \div 9 = 64$

MAD for group Y $=$

$= \dfrac{|58 - 64| + |60 - 64| + |62 - 64| + |63 - 64| + |63 - 64| + |65 - 64| + |67 - 64| + |68 - 64| + |70 - 64|}{9}$

$= \dfrac{6 + 4 + 2 + 1 + 1 + 1 + 3 + 4 + 6}{9} = \dfrac{28}{9} \approx 3.1$

(c) Group Y has a smaller MAD so there is less variability in the data. The data values are clustered around the mean more closely than for Group Y so the mean is a better indicator of a typical value than it is for Group X.

5. Conclusion

Summarize the important points of the lesson.
- We can use MAD to determine whether or not the mean is a good typical value for a set of data.
- When there is low variability (low MAD), the data values are clustered more closely to the mean, so the mean is a good indicator of a typical value.
- When there is high variability (low MAD), the data values are more spread out away from the mean, so the mean is not a good indicator of a typical value.

★ **Workbook: Page 147**

Lesson 11

Objective: Use interquartile range to determine variability around the median.

1. Introduction

Show the scores of two students math tests.

Amaro: 71, 73, 68, 38, 75, 69

Betty: 42, 60, 72, 54, 79

Have students draw a dot plot for each set of data.

Ask students to find the range of each student's scores.

Range of Amaro's scores = 75 − 38 = 37

Range of Betty's scores = 79 − 42 = 37

Ask students to find the mean and Mean Absolute Deviation of each student's scores (rounded the nearest whole number).

Amaro:

- $\dfrac{71 + 73 + 68 + 38 + 75 + 69}{6} = \dfrac{394}{6} \approx 66$

- $\dfrac{|71 - 66| + |73 - 66| + |68 - 66| + |38 - 66| + |75 - 66| + |69 - 66|}{6} = \dfrac{5 + 7 + 2 + 28 + 9 + 3}{6} = \dfrac{54}{6} \approx 9$

Betty:

- $\dfrac{42 + 60 + 72 + 77 + 54 + 69}{6} = \dfrac{384}{6} \approx 64$

- $\dfrac{|41 - 64| + |60 - 64| + |72 - 64| + |77 - 64| + |54 - 64| + |69 - 64|}{6} = \dfrac{23 + 4 + 8 + 13 + 10 + 5}{6} = \dfrac{63}{6} \approx 11$

Ask, "Whose test scores show more variability?"
- Betty's because the MAD is larger.

Ask, "Why do you think their ranges are the same even though Betty's data shows more variability?"
Possible responses:
- Amaro has an extreme value (38) that skews the data.
- The difference between the highest and lowest scores make it seem like they have the same variability but most of Amaro's scores are closer together.

Tell students to remove the extreme value from Amaro's data and find the range, mean, and MAD again.
Amaro's scores (without the extreme value): 71, 73, 68, 75, 69
Range = 75 − 68 = 7

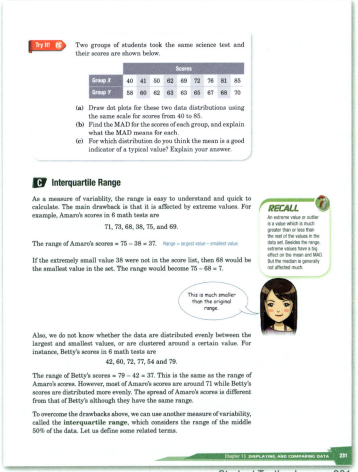

Student Textbook page 231

Amaro:

- $\dfrac{71 + 73 + 68 + 75 + 69}{5} \approx 71$

- $\dfrac{|71 - 71| + |73 - 71| + |68 - 71| + |75 - 71| + |69 - 71|}{5} = \dfrac{0 + 2 + 3 + 4 + 2}{5} = \dfrac{11}{5} \approx 2$

Ask, "How did removing the extreme value affect the MAD?"
- The extreme value made the MAD larger.
- When we take out the extreme value the scores are closer together (less spread out) so the variability is lower.

2. Development

Read and discuss page 231 and the top of page 232.

Notes:
- Outliers in the data can not only skew the mean but they can also skew the range and the MAD.

- Range only takes into account two data points so it is most affected by outliers. Interquartile Range often gives a clearer picture of the true variability of the data by showing the spread of the middle 50% of the data. Since the IQR does not include the lower 25% or upper 25% of the data, the outliers have a smaller effect on the variability.
- IQR is related to median. Three "median" values called **quartiles** divide the data set into four parts that each contain approximately 25% of the values.
  - The first quartile (Q_1) is the median value of the lower half of the data set. It divides the data values into the bottom 25% and the top 75%.
  - The second quartile (Q_2) is the median of the entire data set. It divides the data values into the bottom 50% and the top 50%.
  - The third quartile (Q_3) is the median value of the upper half of the data set. It divides the data values into the bottom 75% and the top 25%.
  - The IQR is the range of values between Q_1 and Q_3.

Students may be confused about why there are three quartiles instead of four. The quartiles are the three "cuts" that divide the data into four sections.

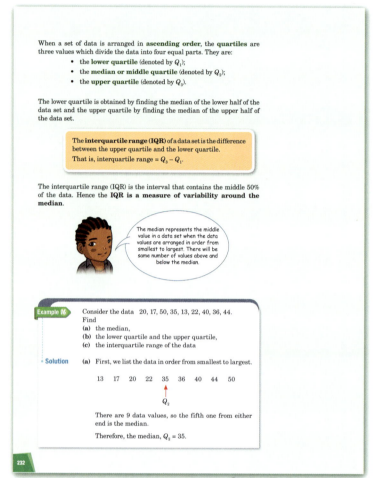

Student Textbook page 232

3. Application

Have students study Example 16 and do Try It! 16.

Notes:

- In Example 15, the range is 50 − 13 = 37. The range makes it seem like the data is more spread out than it really is because of the outliers. The IQR = 23.5 shows the spread of the middle 50% of the data so it is better indicator of the spread of the data.

Try It! 16 Answers

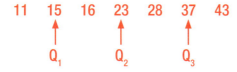

(a) Median = 23

(b) $Q_1 = 15$, $Q_3 = 37$

(c) IQR = 37 − 15 = 22

Lesson 12

4. Extension

Have students study Example 17 and do Try It! 17.

Notes:
- In Example 17, the quartiles are between two values, so we need to find the mean of the two values.
- Have students also compute the range and compare it to the IQR (Range = 41 − 8 = 33, IQR = 20.5). Since range does not take into consideration the outliers, the IQR gives a better picture of the spread.

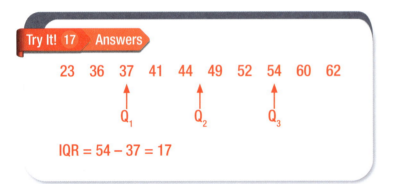

Try It! 17 Answers

23   36   37   41   44   49   52   54   60   62
         Q_1 Q_2 Q_3

IQR = 54 − 37 = 17

5. Conclusion

Summarize the main points of the lesson.
- Extreme values (outliers) can have a big effect on the range and MAD of a set of data.
- To make the effect of the outliers negligible, we can divide the data set into four parts using the median of the lower half of the data set (Q_1), the median of the entire data set (Q_2), and the median of the upper half of the data set (Q_3).
- The difference between Q_3 and Q_1 is the Interquartile Range (IQR). The IQR gives a better picture of the spread of the data because it excludes the values in the upper and lower 25% of the data.

★ **Workbook: Page 151**

Student Textbook page 234

Objective: Use a box plot to summarize a data distribution.

1. Introduction

Discuss how to display the Interquartile Range with a box plot on page 234.

2. Development

Give students grid paper and have them follow the steps for creating a box plot on pages 235 – 236.

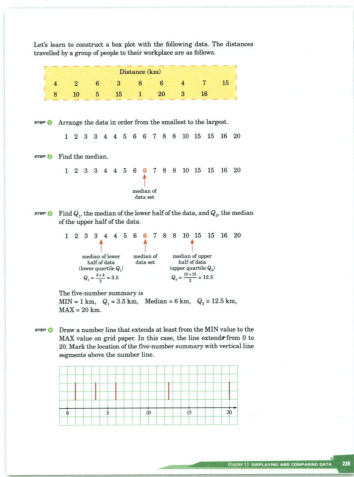

Student Textbook page 235

Discuss the difference between the dot plot and box plot and the merits of each. For example, the bottom of page 236 says, "Unlike the dot plot, a box plot does not show individual data values."

Notes:

- The box plot shows the distribution of data across the range as well as the interquartile range, giving us 5 data points that summarize the variability of the data:
  1. The minimum value of the data set (MIN)
  2. The median of the lower half of the values (Q_1)
  3. The median of the entire data set (Q_2)
  4. The median of the upper half of the values (Q_3)
  5. The maximum value of the data set (MAX)

- The box plot divides the spread of the data into four parts, each containing about 25% of the data values.
  1. The left whisker shows the spread of the lower 25% of the data.
  2. The right whisker shows the spread of the upper 25% of the data.
  3. The box shows the spread of the middle 50% of the data (IQR). The line inside the box is the median of the data set (Q_2). It divides IQR into two parts, each containing half of the data values in the IQR.

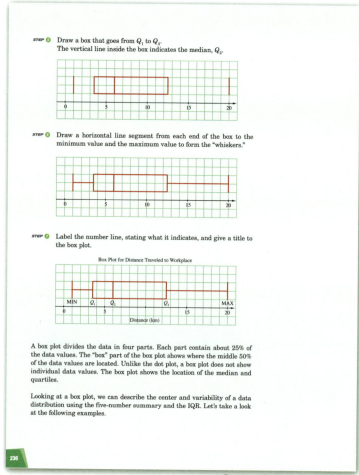

Student Textbook page 236

3. Application

Have students study Example 18 and do Try It! 18.

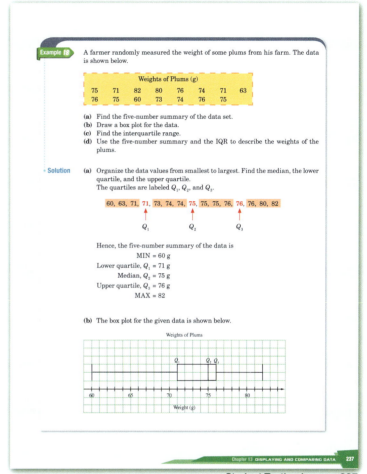

Student Textbook page 237

Notes:

Example 18:
- For 18 (d), the description includes three main points.
    1. The minimum and maximum values of the data set.
    2. The IQR and its meaning (i.e., the values that the middle 50% of the data are between).
    3. A conclusion based on the distribution (e.g., half of the plums weigh more than 75 g).
- Other conclusions can be drawn based on the whiskers. (See REMARK and Note on page 238). The longer the whisker, the greater the spread in that portion of the data. The shorter the whisker, the smaller the spread.

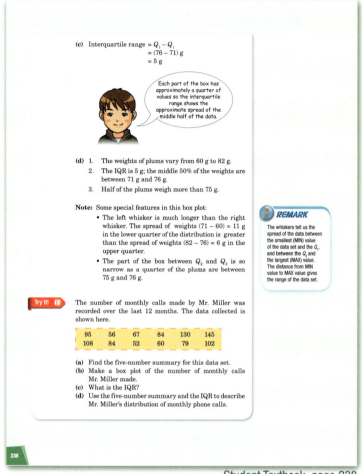

Student Textbook page 238

Try It! 18 Answers

52 56 60 67 79 84 84 95 102 108 130 145

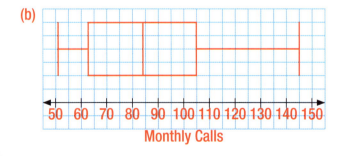

(a) MIN = 52

$Q_1 = \dfrac{60 + 67}{2} = 63.5$

$Q_2 = \dfrac{84 + 84}{2} = 84$

$Q_3 = \dfrac{102 + 108}{2} = 105$

MAX = 145

(b) [box plot on number line from 50 to 150 labeled Monthly Calls]

(c) IQR = 105 − 63.5 = 41.5

(d) 1. The numbers of monthly calls vary from 52 to 145.

   2. The IQR is 41.5; the middle 50% of the numbers of calls are between 63.5 and 105.

   3. Half of the numbers of calls are more than 84.

300

4. Extension

Have students study Example 19 and do Try! It 19.

Notes:

Example 19

- In these problems, we are comparing the variability of two data sets by comparing their box plots.
- The box for Brand *A* is wider than the box for Brand *B*. This means that there is more variability within the middle 50% of the values for Brand *A*. Based on this, it could be concluded that the Brand *A* batteries have less consistent performance than the Brand *B* batteries. (See Note on page 240.)

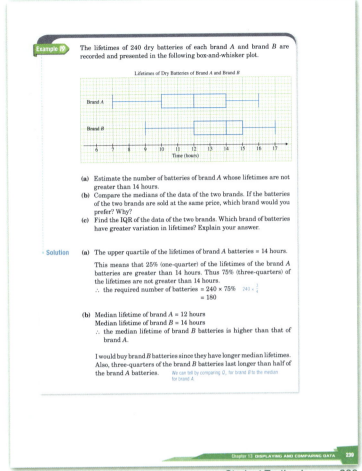

Student Textbook page 239

Try It! 19   Answers

(a) 200 × 75% = 150

(b) Median height of plants under A: 35 cm,
Median heights of plants under B: 39 cm

(c) IQR of A = 45 – 30 = 15 cm
IQR of B = 42 – 31 = 11 cm

(d) Condition B is more favorable because there is a higher median height and less variability.

5. Conclusion

Summarize the main points of the lesson.

- We can use a box plot to get a better picture of the variability of a set of data. The box plot divides the data values into quarters, each containing approximately 25% of the data values.
- The whiskers show the spread of the lower 25% and the upper 25% of the data. The box shows the spread of the middle 50% of the data (the IQR).
- A box plot helps us see variability within the IQR. A wider box means that there is more variability within the middle 50% of the data. A narrower box means there is less variability in the middle 50% of the data.

★ **Workbook: Page 154**

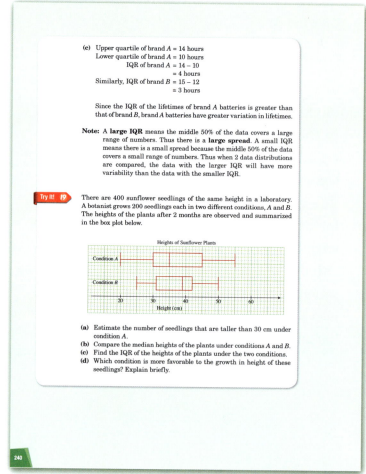

Student Textbook page 240

Lesson 13

Objective: Consolidate and extend the material covered thus far.

Have students work together with a partner or in groups. Students should try to solve the problems by themselves first, then compare solutions with their partner or group. If they are confused, they can discuss together.

Observe students carefully as they work on the problems. Give help as needed individually or in small groups.

BASIC PRACTICE

1. (a) Range: 11, Mean: 26, MAD: 4

    (b) Range: 9, Mean: 4, MAD: 2.8

    (c) Range: 7, Mean: 5, MAD: 2

    (d) Range: 9, Mean 6.5, MAD: 2.25

    (e) Range: 12, Mean: 14, MAD: 3

    Note: In the first printing of the textbook, the answers in the back of the book (page 256) for (b) Mean and MAD and (d) MAD are wrong.

2. (a) $Q_1 = 9$, $Q_2 = 18$, $Q_3 = 24$, IQR = 15

    (b) $Q_1 = 4$, $Q_2 = 5$, $Q_3 = 8.5$, IQR = 4.5

    (c) $Q_1 = 13.5$, $Q_2 = 18$, $Q_3 = 20$, IQR = 6.5

    (d) $Q_1 = 10$, $Q_2 = 20.5$, $Q_3 = 27$, IQR = 17

3. (a) 25 m   (b) 15 m

    (c) 16 m   (d) $Q_1 = 13$ m, $Q_3 = 19$ m

    (e) IQR = 6 m

Student Textbook page 241

4. (a)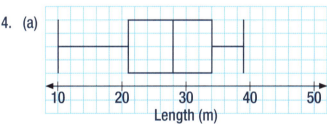

    (b)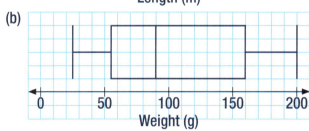

    (c)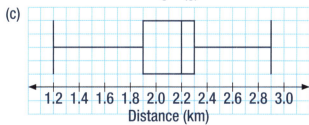

FURTHER PRACTICE

5. (a)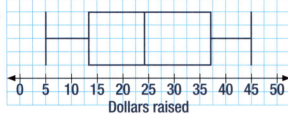

   (b) IQR = 24

   (c) MIN = 5, $Q_1 = 13$, $Q_2 = 24$, $Q_3 = 37$, MAX = 45

6. (a) MIN = 3.4, $Q_1 = 3.5$, $Q_2 = 3.6$, $Q_3 = 3.75$, MAX = 3.9, IQR = 0.25

   (b)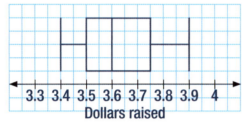

   (c) The distances vary from 3.4 km to 3.9 km.
   The middle 50% of the distances are between 3.5 km and 3.75 km.
   Half of the distances are greater than 3.6 km.

7.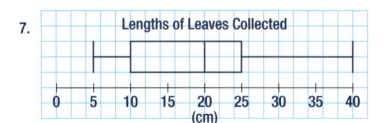

   (a) Range = 40 − 5 = 35 cm

   (b) 25 cm

8. (a) Range: 1.3 m, Mean: 4.5 m, MAD: 0.3 m

   (b) That the variability is low and the jumps are fairly close to the mean.

   (c) The mean and MAD will both decrease.

   (d) No, because the 0 value will skew the mean toward the lower values. Median would be better.

Student Textbook page 242

MATH @ WORK

9. (a) $Q_2 = 70$, IQR = 20, Range = 55

   (b) $Q_2 = 60$, IQR = 25, Range = 60

   (c) Class A, because it has is a higher median.

   (d) Class A, because there is less variability in the IQR.

Note: The answer on textbook page 256 for 9 (d) in the first printing is wrong.

10. School A: 54 60 68 69 73 75 80 81

School B: 58 66 69 69 71 72 72 75

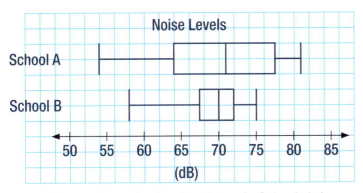

(a) The median noise level is higher in School A. In School A, there is greater variation across the data set and also within the IQR.

(b) School A: Mean: 70 dB, MAD: 7.25 dB.
School B: Mean: 69 dB, MAD: 3.5 dB.

(c) Even though the schools have similar means, the MAD for School A is much higher which means the noise levels for School A have greater variability.

(d) Yes, the predictions in (a) are accurate. Both the MAD and the box plot show greater variability in School A.

11. (a) Same range

(b) There is low variability within the IQR so the middle 50% of wages are close together and close to the median wage.

(c) No. The left whisker and right whisker both show 25% of the wages.

(d) Group 3 has the highest median hourly range. Groups 2 and 1 have the same median hourly wage. The median in Group 3 is near the right end of the box because there is less variability in the wages of the upper half of the box. Half of the wages of the middle 50% of the workers are between $33 and $34 an hour.

(e) Group 1: IQR = $2
Group 2: IQR = $6
Group 3: IQR = $6

(f) Group 1 has the lowest spread in the middle 50%. Groups 2 and 3 have a similar IQR but in Group 3, the middle 50% of the workers have a higher median wage.

12. The IQR would not change.

$$Q_3 - Q_1 = (Q_3 + x) - (Q_1 + x)$$

For example, if Q_3 = \$100,000 and Q_1 = \$75,000, the IQR would be \$100,000 − \$75,000 = \$25,000.

When each staff member's annual income increased by \$3,000:

$$\$100{,}000 - \$75{,}000 = (\$100{,}000 + \$3{,}000) - (\$75{,}000 + \$3{,}000)$$
$$= \$103{,}000 - \$78{,}000$$
$$= \$25{,}000$$

Note: This is based on the properties of subtraction. If you add or subtract the same number to the minuend and subtrahend, the difference will be the same.

$$a - b = (a + c) - (b + c)$$
$$a - b = (a - c) - (b - c)$$

13. Any data set where the sum of the absolute deviation 0 would also have a MAD of 0, so the values would all have to be equal to the mean. For example, for the data set 1, 1, 1, 1, 1:

$$\frac{|1 - 1| + |1 - 1| + |1 - 1| + |1 - 1| + |1 - 1|}{5} = \frac{0}{5} = 0$$

14. (a) Since MAD does not change, that means you added 3 to each of them, so the numbers would be 7, 9, and 14. Algebraically,

$$m = \frac{4 + 6 + 11}{3}$$

$$m + 3 = \frac{(4 + 3) + (6 + 3) + (11 + 3)}{3} = \frac{7 + 9 + 14}{3}$$

The numbers are 7, 9, and 14.

Or,

$$d = \frac{|4 - m + 3| + |6 - m + 3| + |11 - m + 3|}{3}$$

$$= \frac{|7 - m| + |9 - m| + |14 - m|}{3}$$

The numbers are 7, 9, and 14.

(b) Since the MAD is $3d$, that is, the mean absolute deviation is 3 times more, then a result that would give this would be 3 times each data value 4, 6, and 11 so it is 12, 18, and 33. Algebraically,

$$d = \frac{|4 - m| + |6 - m| + |11 - m|}{3}$$

$$3d = 3 \times \frac{|4 - m| + |6 - m| + |11 - m|}{3}$$

$$3d = \frac{3 \times (|4 - m| + |6 - m| + |11 - m|)}{3}$$

$$= \frac{|3 \times 4 - m| + |3 \times 6 - m| + |3 \times 11 - m|}{3}$$

$$= \frac{|12 - 3m| + |18 - 3m| + |33 - 3m|}{3}$$

The numbers are 12, 18, and 33.

Note: The answer in the back of the book (page 256) in the first printing is wrong.

Lesson 14

> **Objective:** Summarize and reflect on important ideas learned in this chapter, and solve a non-routine problem.

Note: This lesson could be done in class or assigned for students to do independently at home or in school.

1. In a Nutshell

Use this page to summarize the important ideas learned in this chapter.

Give examples where needed.

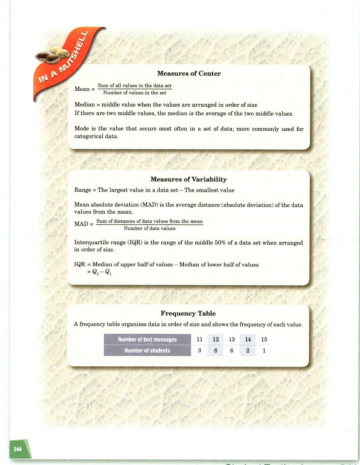

Student Textbook page 244

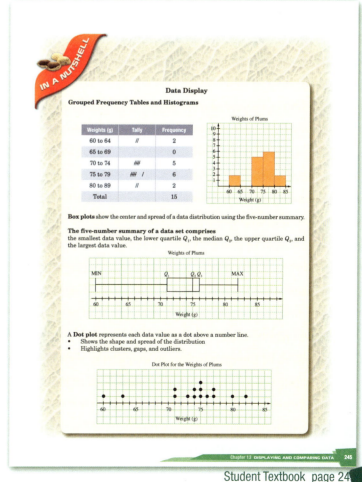

Student Textbook page 245

2. **Write in Your Journal**

Have students complete the writing activity and share their answers. Answers will vary.

Notes:
- For (a), the sum of the values must be 15 because 15 ÷ 3 = 5.
- For (d), the IQR is a better indication when there is a lot of variability in the data. Examples will vary.

3. **Extend Your Learning Curve**

Students can work together with partners or in groups to complete this interesting activity.

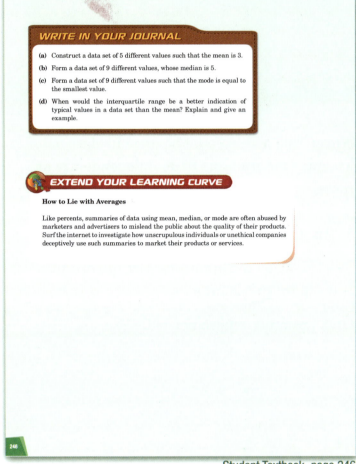

Student Textbook page 246